Hydraulic Fracturing Chemicals and Fluids Technology

T0092071

Hydraulic Fracturing Chemicals and Fluids Technology

Johannes Karl Fink

AMSTERDAM • BOSTON • HEIDELBERG • LONDON
NEW YORK • OXFORD • PARIS • SAN DIEGO
SAN FRANCISCO • SINGAPORE • SYDNEY • TOKYO

Gulf Professional Publishing is an Imprint of Elsevier

Gulf Professional Publishing is an imprint of Elsevier
225 Wyman Street, Waltham, MA 02451, USA
The Boulevard, Langford Lane, Kidlington, Oxford, OX5 1GB, UK

First edition 2013

© 2013 Elsevier Inc. All rights reserved.

No part of this publication may be reproduced, stored in a retrieval system or transmitted in any form or by any means electronic, mechanical, photocopying, recording or otherwise without the prior written permission of the publisher. Permissions may be sought directly from Elsevier's Science & Technology Rights Department in Oxford, UK: phone: (+44) (0) 1865 843830; fax: (+44) (0) 1865 853333; email: permissionselsevier.com. Alternatively visit the Science and Technology website at www.elsevierdirect.com/rights for further information.

Notice

No responsibility is assumed by the publisher for any injury and/or damage to persons or property as a matter of products liability, negligence or otherwise, or from any use or operation of any methods, products, instructions or ideas contained in the material herein. Because of rapid advances in the medical sciences, in particular, independent verification of diagnoses and drug dosages should be made.

Library of Congress Cataloging in Publication Data
A catalog record for this book is available from the Library of Congress

British Library Cataloguing in Publication Data
A catalogue record for this book is available from the British Library

For information on all Gulf Professional Publishing publications
visit our website at *www.elsevierdirect.com*

ISBN: 978-0-12-411491-3

Printed in the United States

Contents

Preface

This manuscript focuses on the recent developments in fracturing fluids with respect to chemical aspects. After a short introduction to the basic issues of fracturing, the text focuses mainly on the organic chemistry of fracturing fluids.

The nature of the individual additives and the justification why the individual additives are acting in the desired way are explained. The material presented here is a compilation from the literature, including patents. In addition, as environmental aspects are gaining increasing importance, this issue is also dealt carefully.

HOW TO USE THIS BOOK

Index

There are four indices, an index of tradenames, an index of acronyms, an index of chemicals, and a general index.

In a chapter, if an acronym is occurring the first time, it is expanded to long form and to short form, e.g., acrylic acid (AA) and placed in the index. If it occurs afterwards it is given in the short form only, i.e., AA. If the term occurs only once in a specific chapter, it is given exclusively in the long form.

In the chemical index, bold face page numbers refer to the sketches of structural formulas or to reaction equations.

Bibliography

The bibliography is given per chapter and is sorted in the order of occurrence. After the bibliography, a list of tradenames that are found in the references and which chemicals are behind these names, as far as laid open, is added.

Acknowledgments

The continuous interest and the promotion by Professor Wolfgang Kern, the head of the department is highly appreciated. I am indebted to our university librarians, Dr. Christian Hasenhttl, Dr. Johann Delanoy, Franz Jurek, Margit Keshmiri, Dolores Knabl, Friedrich Scheer, Christian Slamenik, and Renate Tschabuschnig, for support in literature acquisition. This book could not have been otherwise compiled.

Thanks are given to Professor I. Lakatos, University of Miskolc who directed my interest to this topic.

Last but not least, I want to thank the publisher for kind support, in particular Katie Hammon.

J.K.F.

Chapter 1

General Aspects

Hydraulic fracturing is a technique used to stimulate the productivity of a well. A hydraulic fracture is a superimposed structure that remains undisturbed outside the fracture. Thus the effective permeability of a reservoir remains unchanged by this process.

The increased wellbore radius increases its productivity, because a large contact surface between the well and the reservoir is created.

STRESSES AND FRACTURES

Hydraulic fracturing is one of the newer techniques in petroleum sciences, not being used for more than approximately 60 years. The classic treatment of hydraulic fracturing states that the fractures are approximately perpendicular to the axis of the least stress (Yew, 1997). For most deep reservoirs, the minimal stresses are horizontal, hence vertical stresses will occur in fracturing.

The actual stress can be calculated by balancing the vertical geostatic stress and the horizontal stress by the common tools of the theory of elasticity. For example, the geostatic stress must be corrected in a porous medium filled with a liquid having a poroelastic constant and hydrostatic pressure. The horizontal stress can be calculated from the corrected vertical stress with the *Poisson ratio*. Under some circumstances, in particular in shallow reservoirs, horizontal stresses as well as vertical stresses can be created. The possible stress modes are summarized in Table 1.1.

Fracture Initialization Pressure

Knowledge of the stresses in a reservoir is essential to find the pressure at which initiation of a fracture can take place. The upper bound of this pressure can be estimated using a formula given by von Terzaghi, 1923, which states that

$$p_b = 3s_{H,\min} - s_{H,\max} + T - p. \tag{1.1}$$

The variables are explained in Table 1.1. The closure pressure indicates the pressure at which the width of the fracture becomes zero. This is normally the minimal horizontal stress.

Hydraulic Fracturing Chemicals and Fluids Technology. http://dx.doi.org/10.1016/B978-0-12-411491-3.00001-7
© 2013 Elsevier Inc. All rights reserved.

TABLE 1.1 Modes of Stresses in Fractures

p_b	Fracture initialization pressure
$3s_{H,min}$	Minimal horizontal stress
$s_{H,max}$	Maximal horizontal stress (= minimal horizontal stress + tectonic stress)
T	Tensile strength of rock material
p	Pore pressure

Pressure Decline Analysis

The pressure response during fracturing provides important information about the success of the operation. The fluid efficiency can be estimated from the closure time.

COMPARISON OF STIMULATION TECHNIQUES

In addition to hydraulic fracturing, there are other stimulation techniques, such as acid fracturing or matrix stimulation, and hydraulic fracturing is also used in coal seams to stimulate the flow of methane. Fracturing fluids are often divided into water-based, oil-based, alcohol-based, emulsion, or foam-based fluids. Several reviews are available in the literature dealing with the basic principles of hydraulic fracturing, and the guidelines that are used to select a formulation for a specific job (Ebinger and Hunt, 1989; Ely, 1989; Lemanczyk, 1991).

Polymer hydration, crosslinking, and degradation are the key processes that these materials undergo. Technological improvements over the years have focused primarily on improved rheological performance, thermal stability, and cleanup of crosslinked gels.

Action of a Fracturing Fluid

Fracturing fluids must meet a number of requirements simultaneously. They must be stable at high

- temperatures,
- pumping rates, and
- shear rates.

These severe conditions can cause the fluids to degrade and prematurely settle out the proppant before the fracturing operation is complete. Most commercially used fracturing fluids are aqueous liquids that have been either gelled or foamed.

Typically, the fluids are gelled by a polymeric gelling agent. The thickened or gelled fluid helps keep the proppants within the fluid during the fracturing operation. Fracturing fluids are injected into a subterranean formation for the following purposes (Kelly et al., 2007):

- To create a conductive path from the wellbore extending into the formation.
- To carry proppant material into the fracture to create a conductive path for produced fluids.

Stages in a Fracturing Job

A fracturing job has several stages, including injecting a prepad, a pad, a proppant containing fracturing fluid, and finally, a treatment with flush fluids. A prepad is a low-viscosity fluid used to condition the formation, which may contain fluid loss additives, surfactants, and have a defined salinity to prevent formation damage. The generation of the fractures takes place by injecting the pad, a viscous fluid, but without proppants.

After the fractures develop, a proppant must be injected to keep them permeable. When the fracture closes, the proppant left there creates a large flow area and a highly conductive pathway for hydrocarbons to flow into the wellbore. Thus, the proppant is utilized to maintain an open fracture. Viscous fluids are utilized to transport, suspend, and eventually allow the proppant to be trapped inside the fracture. These fluids typically exhibit a power law behavior for the range of shear rates encountered in hydraulic fracturing treatments.

A uniform proppant distribution is needed in order to get a uniformly conductive fracture along the wellbore height and fracture half-length, but the complicated nature of proppant settling in non-Newtonian fluids often leads to a higher concentration of proppant in the lower part of the fracture. This often leads to a lack of adequate proppant coverage of the upper portion of the fracture and the wellbore. Clustering of proppant, encapsulation, bridging, and embedment are all phenomena that lower the potential conductivity of the proppant pack (Watters et al., 2010).

The job ends eventually with a cleanup stage, in which flush fluids and other cleanup agents are applied. The actual detailed time schedule depends on the particular system used.

After the completion the fluid viscosity should decrease to allow the placement of the proppant and a rapid fluid return through the fracture. It is important to control the time at which the viscosity break occurs. In addition, the degraded polymer should produce little residue to restrict the flow of fluids through the fracture.

SIMULATION METHODS

The estimation of the fracture geometry is one of the most difficult technical challenges in hydraulic fracturing technology (Zhang et al., 2010).

A discrete element simulation for the hydraulic fracture process in a petroleum well that takes into account the elastic behavior of the rock and the *Mohr-Coulomb* fracture criterion has been presented (Torres and Munoz-Castano, 2007). The modeling of the rock was done as an array of Voronoi polygons joined by elastic beams that were submitted to tectonical stresses and the hydrostatic pressure of the fracturing fluid. The fluid pressure is treated similar to that of a hydraulic column. The results show that the simulation process follows a real situation.

A three-dimensional nonlinear fluid solid coupling finite element model was established based on the finite element software ABAQUS® (Zhang et al., 2010). The staged fracturing process of a horizontal well in the Daqing Oilfield, China was simulated with the model, taking account to perforation, wellbore, cement casing, one pay zone, two barriers, micro-annulus fracture, and transverse fracture.

These experimental data were used in numerical computation. Micro-annulus fracture and transverse fracture generate simultaneously and a typical T-shaped fracture occurs at the early stage of treatment history, then the micro-annulus disappears and only the transverse fracture remains and propagates.

The pore pressure distribution in the formation and the fracture configuration during the treatment history can be obtained. The evolution of bottomhole pressure as the direct output of simulation coincides with the experimental data (Zhang et al., 2010).

Productivity

A model has been developed that calculates the productivity of a hydraulically fractured well, including the effect of fracture face damage caused by fluid leakoff (Friehauf et al., 2010).

The results of this model have been compared with those of three previous models. These models assume either elliptical or radial flow around the well, with permeability varying azimuthally. Significant differences in the calculated well productivity indicate that earlier assumptions made regarding the flow geometry can lead to significant overestimates of the well productivity index.

An agreement with the analytical solution given by Levine and Prats, 1961 has been shown for finite-conductivity fractures and no fracture damage.

The simple and discrete nature of the model makes it ideal for implementation in spreadsheets and to connect to fracture performance models. Cleanup of the damage in the invaded zone depends on the capillary properties of the formation and the pressure applied across the damaged zone during production (Friehauf et al., 2010). It has also been found that the invaded zone will cause a significant damage when the permeability of the damaged zone is reduced by more than 9%.

Fracture Propagation

The propagation of a two-dimensional pre-existing fracture in permeable rock by the injection of a viscous, incompressible Newtonian fluid has been modeled (Fareo and Mason, 2010). In particular, the method of Fitt et al., 2007 for the solution of hydraulic fracturing in an impermeable rock has been extended to a permeable rock.

The fluid flow in the fracture is assumed to be laminar. By the application of the lubrication theory, a partial differential equation relating the half-width of the fracture to the fluid pressure and leakoff velocity has been derived. The solution of this equation yields the leakoff velocity as a function of the distance along the fracture and time. The group invariant solution is derived by considering a linear combination of the Lie point symmetries. The boundary value problem has been reformulated as a pair of initial value problems. The model in which the leakoff velocity is proportional to the fracture half-width is considered (Fareo and Mason, 2010).

A model of a hydraulic fracture with a tip cohesive zone according to the Barenblatt approach (Barenblatt et al., 1990) has been presented (Mokryakov, 2011). The particular case of an inviscid fracturing fluid and impermeable rock has been studied. An inviscid fluid is assumed to have zero viscosity. It has been proved that assuming finite cohesive stresses, the cohesive zone length cannot also exceed a certain limit. The limit form of the cohesive zone is obtained, and the limit fracture toughness is thus evaluated. The effective fracture toughness tends to the limit value with power of −0.5.

Proppants

The proppant is the most critical issue of a fracture treatment for a well, as it largely defines the ultimate deliverability. The most efficient and effective fracture stimulation designs create an optimal effective fracture area. This is the area of a created fracture with enough conductivity contrast to promote an accelerated drainage of the reservoir (Brannon and Starks, 2010).

A strong correlation was observed among all the designs between the effective fracture area and predicted cumulative production of 360 d. The cumulative production was significantly less sensitive to the conductivity of the fracture than to the effective area of the fracture. Stimulation designs employing proppant partial monolayers can be competitive or less costly than fracture jobs that use conventional proppants.

Fluid Loss

The modeling of the fluid loss is important for planning a fracturing job and analysis of the finished process. Due to its complexity, theoretical models are set up rather with difficulties, because many of the parameters are difficult or impossible to evaluate.

Empirical and semiempirical models have been set up. Testing protocols also influence the quality of the data on which these models are based. Some of the currently used models have been reviewed and two different models have been proposed that provide better fits to the available data than the previous models (Clark, 2010).

Foam Fluid

A numerical simulation study on the rheology of a foam fluid was carried out by treating the inner phase as a granular fluid using the mixture model (Sun et al., 2012). The results of the simulation show that the gas phase is well distributed. As higher the foam quality, the higher is the velocity gradient near the wall.

In the case of turbulence, both the turbulent kinetic energy and the viscosity increase as the foam quality increases. At a foam quality of 63% the rheology of the foam fracturing fluid changes sharply.

The results of the simulation indicate that the mixture model is more applicable and effective to the region where foam quality is more than 63% (Sun et al., 2012).

Discharge Control

A control and inspection system has been designed which enables the fracture fluid discharge to orderly avoid certain disadvantages such as pressure vibration, sand-back as found sometimes when a manual control is used (Feng and Fu, 2012).

A computer program is used as the interface for conversation between man and machine. The program is designed and used as an online control equipment gathering pressure and flux parameters and control of valve state and status display. In this way the best method of discharging a fracturing fluid is utilized. Actual results have been exemplary presented.

TESTING

Proppant Placement

In hydraulic fracturing treatments, a fracture is initiated by rupturing the formation at high pressure by means of a fracturing fluid. Slurry, composed of propping material carried by the fracturing fluid, is pumped into the induced fracture channel to prevent fracture closure when fluid pressure is released.

The improvement in productivity is mainly determined by the propped dimension of the fracture, which is controlled by the transport of the proppant and a proper placement of the proppant. Settling and convection are the controlling mechanisms of the placement of the proppant. The proppant transport and the placement efficiency using non-Newtonian fluids were experimentally investigated and numerically simulated.

A small glass model was used to simulate the hydraulic fracture and additional parameters, such as slurry volumetric injection rate, proppant concentration, and the rheology properties arising from the polymer (Shokir and Al-Quraishi, 2009). Small glass models can easily and inexpensively simulate the flow patterns in hydraulic fractures. The observed flow patterns are strikingly similar to those obtained by very large flow models.

When the ratio of viscous energy to gravitational energy increases, the settling decreases and the efficiency of proppant placement increases. An enhanced non-Newtonian flow behavior index results in less efficiency of proppant placement.

Slickwater Fracturing

Low-damage fracturing fluids are normally used for better fracture-dimension confinement and lower residue. This leads not only to longer fracture lengths, but also to higher fracture conductivity. The slickwater fracturing technology, developed in the 1980s, is less expensive than gel treatments.

Fluid and proppant volumes can be reduced, and treatment flow rates can be increased significantly. When compared to conventional gel treatments, slickwater fracturing can generate similar or better production responses (Shah and Kamel, 2010)).

Slickwater fracturing has been increasingly applied to stimulate unconventional shale gas reservoirs (Cheng, 2012). In comparison to crosslinked fluids, slickwater used as a fracturing fluid has several advantages, including low cost, a higher possibility of creating complex fracture networks, less formation damage, and ease of cleanup.

An enormous amount of water is injected into the formation during the treatment. Even with a good recovery of injected water from flowback, large quantities of water are still left within the reservoir.

The dynamics of the water phase both within the created hydraulic fractures and the reactivated natural fractures has a significant impact on the effectiveness of the treatment. It is controlled by the relative permeability, capillary pressure, gravity segregation, and the fracture conductivity.

Reservoir-simulation models have been used to investigate the changes of the distribution of the water saturation in the fractures during production. Water imbibitions caused by capillary pressure and gravity segregation can play important roles in distribution of the water saturation, in particular during extended shut-in, which in turn affects the gas flow (Cheng, 2012).

Erosion

Erosion phenomena are common in petroleum industry. Damages of material occur frequently in high-pressure pipelines in the course of hydraulic fracturing (Zhang et al., 2012). With increasing operation times, the erosion and corrosion defects on the inner surface of the pipeline result in serious material loss and thus in a failure of the equipment.

A device to simulate the erosive wear behavior of metal materials caused by a fracturing fluid has been developed (Zhang et al., 2012). The erosion failure mechanism caused by various parameters, such as velocity of multiphase flow, fracturing proppant, and impact angles, has been investigated. Also, microcosmic surface testing was also used to analyze the erosion failure mechanism of metal materials at high-pressure pipelines.

Fluid Leakoff

While a hydraulic fracture is propagating, fluid flow and associated pressure drops must be accounted for both along the fracture path and perpendicularly, into the formation that should be fractured (Economides et al., 2007). Fluid leakoff is the main factor that governs the crack length.

To find an effective approach for the mathematical description of this phenomenon, the thin crack can be represented as the boundary condition for pore pressure spreading in the formation. Such a model was already used for the conduction of heat into a rock formation from a seam in the course of injection of hot water.

A solution for a linearized form of equations has been presented that allows the application of an integral transformation. The resulting analytical solution of the equations includes some integrals that can be calculated numerically.

This model allows a rigorous tracking of the created fracture volume, leakoff volume, and increasing fracture width. The model is advantageous over discreet formulations and allows in time calculations of the resulting fracture dimensions during the injection of the fracturing fluid (Economides et al., 2007).

Damaged Well

A detailed description of the conditions in the hydraulically damaged fracture environment after closure and how to integrate it into a reservoir-simulation model has been presented (Behr et al., 2006).

A special algorithm for the initialization was developed and tested in a support tool to allow the computing of a post-fracture performance in tight gas formations by a reservoir simulator.

To represent the fracture geometry and properties, the information about the distribution of the proppant concentration in the fracture as well as the fracture width variation was translated into the permeabilities and porosities of the fracture gridblocks.

The fracture propagation in the course of the fracturing period under consideration of the leakoff processes was modeled. The penetration of the fracturing fluid into the matrix was modeled by using the Buckley-Leverett equations for a two-phase immiscible displacement (Behr et al., 2006).

Crosslinked Fluids

The rheology and convective heat transfer characteristics of borate crosslinked guar and borate crosslinked foam fracturing fluid have been assessed by conducting experiments on a large-scale test loop at 30 MPa (Sun et al., 2010).

The results indicate a severe chemical degradation of borate crosslinked guar at elevated temperatures. When the temperature was higher than the threshold value, the crosslinking agent was almost disabled and the guar was no longer crosslinked. Also, the viscosity of the borate crosslinked foam fracturing fluid was proportional to the increment of the quality of the foam. This is inversely proportional to the increase in temperature.

The influence of fluid behavior index on the velocity gradient of non-Newtonian fluid at the wall is substantial. A negative temperature-dependent convective heat transfer coefficient is found.

The contribution of shear-induced bubble-scale microconvection is significantly contributing to the heat transfer enhancement of a foam fluid.

Correlations between the viscosity and the convective heat transfer coefficient of the borate crosslinked guar and the foam fracturing fluid were established (Sun et al., 2010).

SPECIAL APPLICATIONS

Unconventional gas reservoirs, including tight gas, shale gas, and coalbed methane, are becoming a critically important component of the current and future gas supply (Lestz et al., 2007).

However, these reservoirs often present unique stimulation challenges. The use of water-based fracturing fluids in low-permeability reservoirs may result in a loss of effective fracture half-length caused by phase trapping associated with the retention of the introduced water into the formation.

This problem is still increased by the water-wet nature of most tight gas reservoirs, because of the strong spreading coefficient of water in such a situation. The retention of increased water saturation in the pore system can restrict the flow of gaseous hydrocarbons, such as methane.

Capillary pressures of 10–20 MPa or higher can be present in low-permeability formations at low water saturation. Furthermore, the use of water in unsaturated reservoirs may also reduce the permeability and thus the gas flow by a continuous increase in water saturation of the reservoir. Compositions of liquid petroleum gas and a volatile hydrocarbon fluid may be helpful for phase trapping of water (Lestz et al., 2007).

Coiled Tubing Fracturing

The success of coiled tubing fracturing in shallow wells has increased the interest in using coiled tubing also for fracturing deeper and hotter wells (Cawiezel et al., 2007).

The key performance requirements of a coiled tubing fracturing fluid for deeper wells are low frictional pressure loss and an adequate proppant-carrying capability after exposure to high-shear zones and high temperatures.

The results of a pilot-scale and field-scale testing have been presented (Cawiezel et al., 2007). These studies resulted in the development of an optimized coiled tubing fracturing fluid.

Polymer-based fracturing fluids can be controllably delayed to have low frictional pressure loss through the curved coiled tubing unit and through a straight tubing. However, the stability of the fluid can be reduced significantly when the fluid is pumped through a tubing with a small diameter and subsequently through high-shear zones.

A correct selection of the fluid is required to meet the requirements in these environments. A successful formulation of a fluid needs to balance the limitations of frictional pressure loss with the maximum rheological stability of the fluid.

Hydrajet Fracturing

Hydrajet fracturing with coiled tubing is a unique technology for low-permeability horizontal and vertical wells. This method uses fluids under high pressure to initiate and accurately place a hydraulic fracture without packer, saving operating time and lowering operating risk.

A hydrajet-fracturing tool has been described in the literature ((Justus, 2007). During the fracturing process the fracturing tool is positioned within a formation to be fractured and fluid is then jetted through the fluid jet against the formation at a pressure sufficient to cut through the casing and cement sheath and form a cavity therein. The pressure must be high enough to also be able to fracture the formation by stagnation pressure in the cavity.

A high stagnation pressure is produced at the tip of a cavity in a formation being fractured because of the jetted fluids being trapped in the cavity as a result of having to flow out of the cavity in a direction generally opposite to the direction of the incoming jetted fluid. The high pressure exerted on the formation at the tip of the cavity causes a fracture to be formed and extend some distance into the formation.

In certain situations, a propping agent is suspended in the fracturing fluid which is deposited in the fracture. The propping agent may be a granular substance such as, for example, sand grains, ceramic or bauxite or other man-made grains, walnut shells, or other material carried in suspension by the fracturing fluid. The propping agent functions to prevent the fractures from closing and thereby provides conductive channels in the formation through which produced fluids can readily flow to the wellbore. The presence of the propping agent also increases the erosive effect of the jetting fluid (Surjaatmadja, 2010).

The mechanisms of hydrajet perforation and hydrajet-fracture initiation have been studied (Tian et al., 2009). Frictions for one kind of fracturing fluid in coiled tubing have been computed to determine the pump pressure and flow rate for field testing.

By comparison of theory and practice it has been proven that the theoretical calculation and field-testing data of hydraulic parameters are basically identical. It has been also proven that the tools meet the requirement of field testing (Tian et al., 2009).

Tight Gas

Hydraulic fracturing is one of the best technologies to improve the productivity from tight gas wells (Gupta et al., 2009). In such low-permeability reservoirs, careful consideration must be given to the proper selection of the fracturing fluid.

Some reservoirs are underpressured and require the use of energized fluids, while others are sensitive to water-based fluids because of clay swelling and migration. Proppant pack damage because of gel residue is one of the primary causes of low production rates after hydraulic-fracturing treatments. In order to minimize the damage and to maximize the production, a new premium, highly efficient fracturing fluid has been developed.

This system is composed of carboxymethyl guar in small amounts and a zirconium crosslinking agent. The delay time of crosslinking can be adjusted, which makes the fluid ideal for deep-well fracturing and coiled tubing treatment since the frictional pressure losses can be minimized. The system can be energized or foamed with carbon dioxide and nitrogen or may also be used in binary foam systems. Case histories have been presented (Gupta et al., 2009).

Energized fluids such as carbon dioxide foam or commingled carbon dioxide fluids are predominantly used for the optimization of the backflow and the proppant transport.

Cold fracturing fluids have a strong cooling effect in the fracturing process mainly in rock-stress reduction, however these kinds of fluids should be precisely evaluated, in terms of thermal stress effect, cleanup process, and proppant movement (Rafiee et al., 2009). A cold water fluid has been tested successfully in geothermal reservoirs. This stimulation technique is successful in these fields but not in tight gas reservoirs.

The use of liquid carbon dioxide as a fracturing fluid offers a viable method of stimulation. The successful application of these fluids to a variety of formations in USA was shown. The process has proven to be an economical alternative to conventional stimulation fluids.

Liquid nitrogen can also be used in limited cases. From those results it can be concluded that liquid carbon dioxide seems to be the most efficient fracturing fluid for tight gas stimulations.

The coupling of the cold fracture technology with underground carbon dioxide capture and storage is an interesting promising alternative possibility (Rafiee et al., 2009).

Shale Gas

A method has been described for the in situ production of oil shale and gas hydrates wherein a network of fractures is formed by injecting liquified gases into a horizontally disposed fracturing borehole. Heat is thereafter applied to liquify the kerogen or to dissociate the gas hydrates so that oil shale oil or gases can be recovered from the fractured formations (Maguire, 2007).

Liquid nitrogen injected at a fracturing pressure of 500 psi will increase its volume 14-fold at a temperature of −60 °C (−75 °F). If, however, no increase in fracture volume occurs, the expansion pressure would increase to approximately 7000 psia at this temperature.

The utilization of water as a heating agent is important because the injection of the water will replace the void spaces created by the dissociation of the gas hydrate and the shrinking of the hydrate ice and will also prevent possible slumping of the hydrate beds (Maguire, 2010).

After injection, the heated water will dissociate the gas hydrate and the gas will migrate downward through the created fracture system to the lower production borehole and into the casing annulus and thence to the surface.

The required heat to heat the water can be supplied by the combustion of the produced gas in fuel steam generators. The injection of the hot water should occur into the top of the gas hydrate zone to permit the injected water to migrate downward so that no old water would steal heat from the new water being injected as would occur if injection was instigated from the lower zone.

The hydrostatic pressure and increased injection pressure, with a lower production borehole pressure, would force the liberated gas to flow downward rather than upward from the buoyancy factor (Maguire, 2010).

Coalbed Methane

The production of natural gas from coal typically requires a stimulation with hydraulic fracturing. Basic studies on the effectiveness of various treatment methods for coalbeds have been presented in the literature (Conway and Schraufnagel, 1995; Penny and Conway, 1995).

Treating a coal seam with a well treatment fluid containing a dewatering agent will enhance the methane production through a well. This additive enhances the permeability of the formation to water production and binds tenaciously to the coal surface so that the permeability-enhancement benefits are realized over a long production term.

Dewatering surfactants can be poly(oxyethylene), poly(oxypropylene), and poly(ethylene carbonate)s (Nimerick and Hinkel, 1991) or *p-tert*-amylphenol condensed with formaldehyde, or they can be composed of a copolymer from 80% to 100% alkyl methacrylate monomers and hydrophilic monomers (Harms and Scott, 1993). Selected compounds for this purpose are shown in Figure 1.1.

Such a well treatment fluid may be used in both fracturing and competition operations to enhance and maintain fracture conductivity over an extended period of production.

Active water and plant gum fracturing fluids are widely used for coalbed methane in China (Dai et al., 2011). However, there are limitations due to poor rheological properties, high content of water-insoluble substances, and high residual content.

A gel fracturing fluid was used that is composed of nonionic poly(acrylamide), $ZrOCl_2$ as crosslinking agent, a pH modifier, a gel breaker with $(NH_4)_2S_2O_8$, and an activator. This formulation was used in a low-temperature (20–40 °C) and low-permeability coalbed methane reservoir.

This type of gel fracturing fluid has the advantages of easy preparation, low cost, strong shearing resistance capacity, low filtration coefficient, rapid

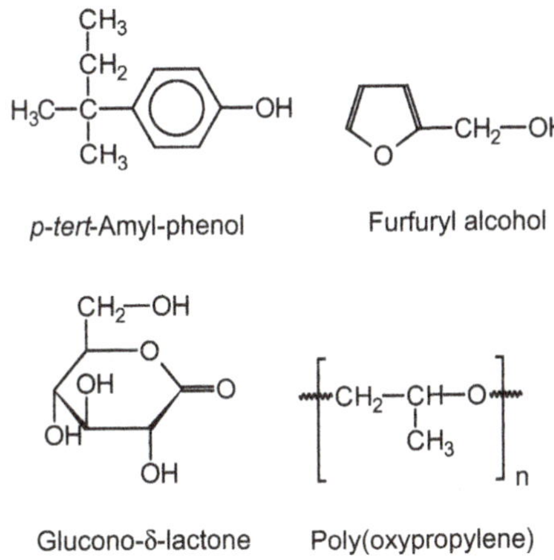

p-tert-Amyl-phenol

Furfuryl alcohol

Glucono-δ-lactone

Poly(oxypropylene)

FIGURE 1.1 Monomers for dewatering.

break, lack of residual after gel breaking, and ease of flowback. Further, the performance of gel fracturing fluid is superior to those of active water and vegetable gum, and it is very suitable for use as a fracturing treatment in coalbed methane at low temperatures Dai et al., 2011).

REFERENCES

Barenblatt, G.I., Entov, V.M., Ryzhik, V.M., 1990. Theory of Fluid Flows Through Natural Rocks. Kluwer Academic Publishers, Dordrecht, Boston.

Behr, A., Mtchedlishvili, G., Friedel, T., Haefner, F., 2006. Consideration of damaged zone in a tight gas reservoir model with a hydraulically fractured well. SPE Prod. Oper. 21 (2), 206–211. http://dx.doi.org/10.2118/82298-PA.

Brannon, H., Starks, T.R.I.I., 2010. Less can deliver more. Oilfield Technol. 3 (2), 59–63.

Cawiezel, K.E., Wheeler, R.S., Vaughn, D.R., 2007. Specific fluid requirements for successful coiled-tubing fracturing applications. SPE Prod. Oper. 22 (1), 83–93. http://dx.doi.org/10.2118/86481-PA.

Cheng, Y., 2012. Impact of water dynamics in fractures on the performance of hydraulically fractured wells in gas-shale reservoirs. J. Can. Petrol. Technol. 51 (2), 143–151.

Clark, P.E., 2010. Analysis of fluid loss data II: models for dynamic fluid loss. J. Petrol. Sci. Eng. 70 (3–4), 191–197. http://dx.doi.org/10.1016/j.petrol.2009.11.010.

Conway, M.W., Schraufnagel, R.A., 1995. The effect of fracturing fluid damage on production from hydraulically fractured wells. In: Proceedings Volume. Alabama University et al International Unconventional Gas Symposium (Intergas 95) (Tuscaloosa, AL, 14–20 May 1995), pp. 229–236.

Dai, C., You, Q., Zhao, H., Guan, B., Wang, X., Zhao, F., 2011. A study on gel fracturing fluid for coalbed methane at low temperatures. Energy Sources Part A: Recovery Utiliz. Environ. Effects 34 (1), 82–89. http://dx.doi.org/10.1080/15567036.2010.545806.

Ebinger, C.D., Hunt, E., 1989. Keys to good fracturing: Pt. 6: new fluids help increase effectiveness of hydraulic fracturing. Oil Gas J. 87 (23), 52–55.

Economides, M.J., Mikhailov, D.N., Nikolaevskiy, V.N., 2007. On the problem of fluid leakoff during hydraulic fracturing. Transport Porous Med. 67 (3), 487–499. http://dx.doi.org/10.1007/s11242-006-9038-7.

Ely, J.W., 1989. Fracturing fluids and additives. In: Henry L. Doherty (Ed.), Recent Advances in Hydraulic Fracturing, vol. 12 (SPE Monogr Ser). SPE, Richardson, Texas.

Fareo, A.G., Mason, D.P., 2010. A group invariant solution for a pre-existing fluid-driven fracture in permeable rock. Nonlinear Anal.: Real World Appl. 12 (1), 767–779. http://dx.doi.org/10.1016/j.nonrwa.2010.08.004.

Feng, Y., Fu, K., 2012. Automatic control of liquid discharging after hydraulic fracture of oil well. Adv. Mater. Res. 443–444 (Pt. 2, Manufacturing Science and Materials Engineering), 774–778. http://dx.doi.org/10.4028/www.scientific.net/AMR.443-444.774.

Fitt, A.D., Mason, D.P., Moss, E.A., 2007. Group invariant solution for a pre-existing fluid-driven fracture in impermeable rock. Zeitschrift für angewandte Mathematik und Physik 58 (6), 1049–1067. http://dx.doi.org/10.1007/s00033-007-7038-2.

Friehauf, K.E., Suri, A., Sharma, M.M., 2010. A simple and accurate model for well productivity for hydraulically fractured wells. SPE Prod. Oper. 25 (4), 453–460.

Gupta, D.V.S., Jackson, T.L., Hlavinka, G.J., Evans, J.B., Le, H.V., Batrashkin, A., Shaefer, M.T., 2009. Development and field application of a low-pH, efficient fracturing fluid for tight gas fields in the Greater Green River basin, Wyoming. SPE Prod. Oper. 24 (4), 602–610. http://dx.doi.org/10.2118/116191-PA.

Harms, W.M., Scott, E., 1993. Method for stimulating methane production from coal seams. US Patent 5 249 627, assigned to Halliburton Co., October 5 1993.

Justus, D.M., 2007. Hydrajet perforation and fracturing tool. US Patent 7 159 660, assigned to Halliburton Energy Services, Inc. (Duncan, OK), January 9 2007. <http://www.freepatentsonline.com/7159660.html>.

Kelly, P.A., Gabrysch, A.D., Horner, D.N., 2007. Stabilizing crosslinked polymer guars and modified guar derivatives. US Patent 7 195 065, assigned to Baker Hughes Incorporated (Houston, TX), March 27 2007. <http://www.freepatentsonline.com/7195065.html>.

Lemanczyk, Z.R., 1991. The use of polymers in well stimulation: performance, availability and economics. In: Proceedings Volume. Plast Rubber Institute Use of Polymers in Drilling & Oilfield Fluids Conference (London, England, 12 September 1991).

Lestz, R.S., Wilson, L., Taylor, R.S., Funkhouser, G.P., Watkins, H., Attaway, D., 2007. Liquid petroleum gas fracturing fluids for unconventional gas reservoirs. J. Can. Petrol. Technol. 46 (12), 68–72. http://dx.doi.org/10.2118/07-12-03.

Levine, J.S., Prats, M., 1961. The calculated performance of solution-gas-drive reservoirs. Soc. Petrol. Eng. J. 1 (3), 142–152. http://dx.doi.org/10.2118/1520-G.

Maguire, J.Q., 2007. In-situ method of producing oil shale and gas (methane) hydrates, on-shore and off-shore. US Patent 7 198 107, assigned to Maguire and James Q. (Norman, OK), April 3 2007. <http://www.freepatentsonline.com/7198107.html>.

Maguire, J.Q., 2010. In-situ method of fracturing gas shale and geothermal areas. US Patent 7 784 545, August 31 2010. <http://www.freepatentsonline.com/7784545.html>.

Mokryakov, V., 2011. Analytical solution for propagation of hydraulic fracture with barenblatt's cohesive tip zone. Int. J. Fracture 169 (2), 159–168. http://dx.doi.org/10.1007/s10704-011-9591-0.

Nimerick, K.H., Hinkel, J.J., 1991. Enhanced methane production from coal seams by dewatering. EP Patent 444 760, assigned to Pumptech NV and Dowell Schlumberger SA, September 4 1991.

Penny, G.S., Conway, M.W., 1995. Coordinated studies in support of hydraulic fracturing of coalbed methane: Final report (July 1990–May 1995). Gas Res Inst Rep GRI-95/0283, Gas Res Inst.

Rafiee, M.M., Wilsnack, T., Voigt, H.D., Haefner, F., 2009. Cold-frac technology in tight gas reservoirs. DGMK Tagungsbericht 1 (DGMK/OGEW-Frnhjahrstagung des Fachbereiches Aufsuchung und Gewinnung, 2009), 441–450.

Shah, S.N., Kamel, A.H.A., 2010. Investigation of flow behavior of slickwater in large straight and coiled tubing. SPE Prod. Oper. 25 (1), 70–79. http://dx.doi.org/10.2118/118949-PA.

Shokir, E.M., Al-Quraishi, A.A., 2009. Experimental and numerical investigation of proppant placement in hydraulic fractures. Petrol. Sci. Technol. 27 (15), 1690–1703. http://dx.doi.org/10.1080/10916460802608768.

Sun, X., Wang, S., Lu, Y., 2012. Study on the rheology of foam fracturing fluid with mixture model. Adv. Mater. Res. 512–515 (Pt. 3, Renewable and Sustainable Energy II), 1747–1752. http://dx.doi.org/10.4028/www.scientific.net/AMR.512-515.1747.

Sun, X., Wang, S.-Z., Bai, Y., Liang, S.-S., 2010. Rheology and convective heat transfer properties of borate cross-linked nitrogen foam fracturing fluid. Heat Transfer Eng. 32 (1), 69–79. http://dx.doi.org/10.1080/01457631003732979.

Surjaatmadja, J.B., 2010. Hydrajet tool for ultra high erosive environment. US Patent 7 841 396, assigned to Halliburton Energy Services Inc. (Duncan, OK), November 30 2010. <http://www.freepatentsonline.com/7841396.html>.

Tian, S., Li, G., Huang, Z., Niu, J., Xia, Q., 2009. Investigation and application for multistage hydrajet-fracturing with coiled tubing. Petrol. Sci. Technol. 27 (13), 1494–1502. http://dx.doi.org/10.1080/10916460802637569.

Torres, G.S.A., Munoz-Castano, J.D., 2007. Simulation of the hydraulic fracture process in two dimensions using a discrete element method. Physical Review E: Statistical, Nonlinear Soft Matter Phys. 75 (6–2), 066109/1–066109/9. http://dx.doi.org/10.1103/PhysRevE.75.066109.

von Terzaghi, K., 1923. Die Berechnung der Durchlässigkeitsziffer des Tones aus dem Verlauf der hydrodynamischen Spannungserscheinungen. Sitzungsberichte der Akademie der Wissenschaften in Wien, Mathematisch-Naturwissenschaftliche Klasse, Abteilung 2a.

Watters, J.T., Ammachathram, M., Watters, L.T., 2010. Method to enhance proppant conductivity from hydraulically fractured wells. US Patent 7 708 069, assigned to Superior Energy Services, L.L.C. (New Orleans, LA), May 4 2010. <http://www.freepatentsonline.com/7708069.html>.

Yew, Ching H., 1997. Mechanics of Hydraulic Fracturing. Gulf Pub. Co., Houston, Tex. http://www.worldcat.org/title/mechanics-of-hydraulic-fracturing/oclc/162129743&referer=brief_results.
Zhang, G.M., Liu, H., Zhang, J., Wu, H.A., Wang, X.X., 2010. Three-dimensional finite element simulation and parametric study for horizontal well hydraulic fracture. J. Petrol. Sci. Eng. 72 (3–4), 310–317. http://dx.doi.org/10.1016/j.petrol.2010.03.032.
Zhang, J., Fan, J., Xie, Y., Wu, H., 2012. Research on erosion of metal materials for high pressure pipelines. Adv. Mater. Res. 482–484 (2), 1592–1595. http://dx.doi.org/10.4028/www.scientific.net/AMR.482-484.1592.

Chapter 2

Fluid Types

Generally, a hydraulic fracturing treatment involves pumping a proppant-free viscous fluid, or pad, which is usually water with some fluid additives, in order to generate high viscosity, into a well faster than the fluid can escape into the formation. This causes the pressure to rise and the rock to break, creating artificial fractures or enlarging existing ones.

After fracturing the formation, a propping agent, such as sand, is added to the fluid. This forms a slurry that is pumped into the newly formed fractures in the formation to prevent them from closing when the pumping pressure is released. The proppant transportability of a base fluid depends on the type of viscosifying additives that have been added to the water base (Lukocs et al., 2007).

Since the late 1950s, more than half of fracturing treatments have been conducted with fluids comprising guar gums, or guar derivatives such as hydropropyl guar (HPG), hydroxypropyl cellulose (HPC), carboxymethyl guar, and carboxymethyl hydropropyl guar.

Crosslinking agents based on boron, titanium, zirconium, or aluminum complexes are typically used to increase the effective molecular weight of the polymers and make them better suited for use in high-temperature wells.

Cellulose derivatives, such as hydroxyethyl cellulose (HEC) or HPC and carboxymethylhydroxyethyl cellulose, are also used, with or without crosslinkers. Xanthan and scleroglucan have also been shown to have excellent proppant-suspension ability, but they are more expensive than guar derivatives and therefore are used less frequently.

Poly(acrylamide) (PAM) and polyacrylate polymers and copolymers are typically used for high-temperature applications or as friction reducers at low concentrations for all temperatures ranges (Lukocs et al., 2007).

Polymer-free, water-based fracturing fluids can be obtained by using viscoelastic surfactants (VES)s. These fluids are normally prepared by mixing appropriate amounts of suitable surfactants such as anionic, cationic, nonionic, and zwitterionic surfactants. Their viscosity is attributed to the three-dimensional structure formed by the components in the fluids. The viscosity increases when the surfactant concentration exceeds a critical concentration. Then the surfactant molecules aggregate into micelles, which can interact to form a network that exhibits viscous and elastic behavior.

Hydraulic Fracturing Chemicals and Fluids Technology. http://dx.doi.org/10.1016/B978-0-12-411491-3.00002-9
© 2013 Elsevier Inc. All rights reserved.

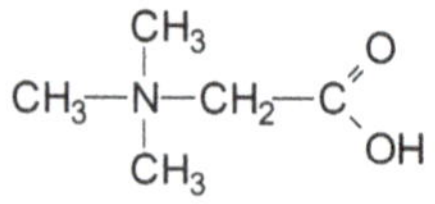

$$\begin{array}{c} CH_3 \\ | \\ CH_3{-}N{-}CH_2{-}C(=O){-}OH \\ | \\ CH_3 \end{array}$$

FIGURE 2.1 Betaine.

Cationic VESs—typically consisting of long-chain quaternary ammonium salts such as cetyltrimethylammonium bromide—have so far been the type attracting most commercial interest. Other common reagents that generate viscoelasticity in surfactant solutions include salts, such as ammonium chloride, potassium chloride, sodium chloride, sodium salicylate, and sodium isocyanate, and nonionic organic molecules, such as chloroform. The electrolyte content of surfactant solutions is also important for controlling their viscoelastic behavior (Lukocs et al., 2007).

Fluids of this type of cationic VESs tend to lose their viscosity at high brine concentrations, hence they have seen limited use as gravel packing or drilling fluids. Anionic VESs are also used.

Amphoteric/zwitterionic surfactants (Allan et al., 2008) and an organic acid, salt, or an inorganic salt can also impart viscoelastic properties. The surfactants could be, for dihydroxyl alkyl glycinate, alkyl amphoacetate or VESs propionate, alkyl betaine, alkyl amidopropyl betaine, and alkylamino mono- or dipropionates derived from certain waxes, fats, and oils. They are used in conjunction with an inorganic water-soluble salt or organic additives such as phthalic acid, salicylic acid, or their salts.

Amphoteric/zwitterionic surfactants, in particular those comprising a betaine moiety, are useful for a temperature up to about 150 °C and are therefore of particular interest for medium- to high-temperature wells. Betaine is shown in Figure 2.1. However, like the cationic viscoelastic surfactants mentioned above, anionic surfactants are usually not compatible with high brine concentration.

Polymer-free VES fluids are used to minimize the damage to the proppant pack and to efficiently transport the proppants into fractures. The proper assessment of the rheologic properties and the proppant settling of the fluids plays an important role in fracturing engineering (Wang et al., 2012).

The rheology and viscosity-temperature properties of a VES fracturing fluid have been investigated. The VES fluid behaves as a non-Newtonian shear thinning fluid and the power law model can be used to describe fluid rheology within a certain range of shear rate and temperature. However, with an increase of the shear rate and the temperature, the fluid approaches a Newtonian fluid.

When the VES concentration is 4%, the fluid may generate a stable micro-mesh wormlike micelle structure, which results in a good viscoelasticity and a high proppant-carrying capacity (Wang et al., 2012).

A formulation of an easy-to-prepare, surfactant-based polymer-free fluid has been described (Deng et al., 2012). Possible components in the fluid that have been tested are tetradecyl trimethyl ammonium bromide, cetyl trimethyl ammonium bromide, octadecyl trimethyl ammonium bromide, and salicylic acid. The properties as a fracturing fluid were evaluated with regard to viscoelasticity and proppant-carrying capability.

These water-based gels have a strong capacity to carry proppants and a high viscoelasticity. The best performance was obtained from the gels derived from octadecyl trimethyl ammonium bromide with salicylic acid. The viscoelasticity of these gels increases with the ratio of ammonium bromide to acid (Deng et al., 2012).

Proppants can be sand, intermediate strength ceramic proppants, or sintered bauxites, which can be coated with a resin to improve their clustering ability. They can be coated with resin or a proppant flowback control agent such as fibers. By selecting proppants having a contrast in property such as density, size, or concentration, different settling rates will be achieved.

Waterfrac treatments combine low cost, low viscosity fluids to stimulate very low-permeability reservoirs. The treatments rely on the mechanisms of asperity creation (rock spalling), shear displacement of rock, and localized high concentration of proppant to create adequate conductivity, with the last mechanism being mostly responsible for the success of the treatment. The mechanism can be described as analogous to a wedge splitting wood.

A viscous well treatment fluid is generally composed of a polysaccharide or synthetic polymer in an aqueous solution, which is crosslinked by an organometallic compound. Examples of well treatments in which metal crosslinked polymers are used are hydraulic fracturing, gravel packing operations, water blocking, and other well completion operations.

In order for the treatment to be successful, the fluid viscosity should eventually diminish to levels approaching that of water after the proppant is placed. This allows a portion of the treating fluid to be recovered without producing excessive amounts of proppant after the well is opened and returned to production. If the viscosity of the fluid is low, it will flow naturally from the formation under the influence of formation fluids. This viscosity reduction or conversion is referred to as breaking, and is accomplished by incorporating chemical agents, referred to as breakers, into the initial gel.

Some fracturing fluids, such as those based upon guar polymers, break naturally without the intervention of a breaking agent, but their breaking time is generally somewhere in the range from greater than 24 h, to weeks, months, or years depending on the conditions in the reservoir.

To decrease this break time, chemical agents are usually incorporated into the gel. These are typically either oxidants or enzymes that degrade the polymeric gel structure. Oxidizing agents, such as persulfate salts, chromous salts, organic peroxides or alkaline earth or zinc peroxide salts, or enzymes are the most effective.

The timing of the break is also of great importance. Gels that break prematurely can cause suspended proppant material to settle out. They penetrate before a sufficient distance into the produced fracture. Premature breaking can also lead to a premature reduction in the fluid viscosity, resulting in an inadequate fracture width. On the other hand, gelled fluids that break too slowly can cause slow recovery of the fracturing fluid, with attendant delay in resuming production.

Additional problems may occur, such as the tendency of proppant to become dislodged from the fracture, resulting in at least partial closing and decreased efficiency of the fracturing operation. The fracturing gel should preferably begin to break when the pumping operations are finished, and be completely broken within about 24 h after completion of the treatment.

Fracturing fluid compositions comprise a solvent, a polymer-soluble or hydratable in the solvent, a crosslinking agent, an inorganic breaking agent, an optional ester compound, and a choline carboxylate. The solvent may be an aqueous potassium chloride solution, and the inorganic breaking agent may be a metal-based oxidizing agent, such as an alkaline earth metal or a transition metal, or it may be magnesium, calcium, or zinc peroxide. The ester compound may be an ester of a polycarboxylic acid, such as an ester of oxalate, citrate, or ethylenediamine tetraacetate. Those having hydroxyl groups can also be acetylated, for instance, citric acid can be acetylated to form acetyl triethyl citrate, which is a preferred ester.

The hydratable polymer can be a water-soluble polysaccharide, such as galactomannan or cellulose, and the crosslinking agent may be a borate, titanate, or zirconium-containing compound, such as $Na_3BO_3 \times nH_2O$.

A general review of commercially available additives for fracturing fluids is given in the literature (Anonymous, 1999). Possible components in a fracturing fluid are listed in Table 2.1, which indicates the complexity of a fracturing fluid formulation. Some additives may not be used together, such as oil-gelling additives in a water-based system. More than 90% of the fluids are based on water. Aqueous fluids are economical and can provide control of a broad range of physical properties if used with additives. Additives for fracturing fluids serve two purposes (Harris, 1988):

1. They enhance fracture creation and proppant-carrying capability.
2. They minimize formation damage.

Viscosifiers, such as polymers and crosslinking agents, temperature stabilizers, pH control agents, and fluid loss control materials, assist the creation of a fracture. Formation damage is reduced by gel breakers, biocides, surfactants, clay stabilizers, and gases. Table 2.2 summarizes the various types of fluids and techniques used in hydraulic fracturing.

COMPARISON OF DIFFERENT TECHNIQUES

The optimal technique to be used depends on the type of reservoir. Reports that compare the techniques in a related environment are available. In the Kansas Hugoton field (Mesa Limited Partnership), several hydraulic fracturing methods were tested (Cottrell et al., 1988).

TABLE 2.1 Components in Fracturing Fluids

Component/category	Function/remark
Water-based polymers	Thickener, to transport proppant, reduces leakoff in formation
Friction reducers	Reduce drag in tubing
Fluid loss additives	Form filter cake, reduce leakoff in formation if thickener is not sufficient
Breakers	Degrade thickener after job or disable cross-linker (wide variety of chemical mechanisms)
Emulsifiers	For diesel premixed gels
Clay stabilizers	For clay-bearing formations
Surfactants	Prevent water-wetting of formation
Nonemulsifiers	Destroy emulsions
pH-Control additives	Increase the stability of fluid (e.g., for elevated temperature applications)
Crosslinkers	Increase the viscosity of the thickener
Foamers	For foam-based fracturing fluids
Gel stabilizers	Keep gels active longer
Defoamers	Break a foam
Oil-gelling additives	Same as crosslinkers for oil-based fracturing fluids
Biocides	Prevent microbial degradation
Water-based gel systems	Common
Crosslinked gel systems	Increase viscosity
Oil-based systems	Used in water sensitive formation
Polymer plugs	Used also for other operations
Gel concentrates	Premixed gel on diesel base
Resin-coated proppants	Proppant material
Ceramics	Proppant material

A method in which a complexed gelled water fracture was applied was the most successful when compared with a foam technique and with older and simpler techniques. The study covers some 56 wells where such techniques were applied.

EXPERT SYSTEMS FOR ASSESSMENT

A PC-based, interactive computer model has been developed to help engineers choose the best fluid and additives and the most suitable propping agent for a given set of reservoir properties (Holditch et al., 1993; Xiong et al., 1996).

TABLE 2.2 Various Types of Hydraulic Fracturing Fluids

Type	Remarks
Water-based fluids	Predominant
Oil-based fluids	Water sensitive; increased fire hazard
Alcohol-based fluids	Rarely used
Emulsion fluids	High pressure, low temperature
Foam-based fluids	Low pressure, low temperature
Noncomplex gelled water fracture	Simple technology
Nitrogen-foam fracture	Rapid cleanup
Complexed gelled water fracture	Often the best solution
Premixed gel concentrates	Improve process logistics
In situ precipitation technique	Reduces scale-forming ingredients (Hrachovy (1994a,1994b))

The model also optimizes treatment volume, based on reservoir performance and economics. To select the fluids, additives, and propping agents, the expert system surveys stimulation experts from different companies, reviews the literature, and then incorporates the knowledge so gained into rules, using an expert system shell.

In addition, the fluid leakoff during hydraulic fracturing can be modeled, calculated, and measured experimentally. Procedures for converting laboratory data to an estimate of the leakoff under field conditions have been given in the literature (Penny and Conway 1989).

OIL-BASED SYSTEMS

One advantage of fracturing with hydrocarbon gels compared with water-based gels is that some formations may tend to imbibe large quantities of water, whereas others are water sensitive and will swell if water is introduced.

FOAM-BASED FRACTURING FLUIDS

Foam fluids can be used in many fracturing jobs, especially when environmental sensitivity is a concern (Stacy and Weber, 1995). Foam-fluid formulations are reusable, are shear stable, and form stable foams over a wide temperature range. They exhibit high viscosities even at relatively high temperatures (Bonekamp et al., 1993).

Carbon dioxide, nitrogen, and binary high quality foams are widely used in tight and deep formations due to their capacity to energize the fluid and improve the total flowback volume and rate (Tamayo et al., 2008).

A foamed fracturing fluid has a relatively large volume of gas dispersed in a relatively small volume of liquid. A foamed fracturing fluid also includes a surfactant for facilitating the foaming and stabilization of the foam produced when the gas is mixed with the liquid (Welton et al., 2010).

A coarse foamed fluid has a relatively nonuniform bubble size distribution, e.g., a combination of large and small gas bubbles, whereas a fine texture foam has relatively uniform bubble size distribution and most of the bubbles are relatively small (Middaugh et al., 2007).

In coarse foamed fracturing fluids, there may be also regions of fine textured foam. Such foams are able to support the proppant in the fine textured regions even at very high foam quality levels. The most commonly used gases for foamed fracturing fluids are nitrogen and carbon dioxide because they are noncombustible, readily available, and relatively cheap (Welton et al., 2010).

Surfactants designed to reduce surface and interfacial tension are also a key element in the design of fluid systems to enhance recovery and reduce entrapment of fluid barriers within the formation (Tamayo et al., 2008). Enhanced fluid recovery improves overall completions economics due to less total treatment cost and less time required for flowing back fluids. The most important benefit is achieving a less damaged proppant pack, resulting in higher fracture conductivity.

The application of carbon dioxide foamed fluids and surfactants to enhance fracturing fluid recovery has been reviewed (Tamayo et al., 2008).

Recyclable foamed fracturing fluids are available (Chatterji et al., 2007). After placement, the pH of the fracturing fluid is changed so that the foam is destroyed. At this stage, the fracturing fluid also releases its proppant. Afterwards, the fracturing fluid is allowed to flow back to the surface. Eventually, it can be recycled by changing the pH of the fracturing fluid back to the first pH and adding a gas to the fluid, causing it to foam again.

ACID FRACTURING

A difference exists between acid fracturing and matrix acidizing. Acid fracturing is used for low-permeability, acid-soluble rocks. Matrix acidizing is a technique used for high-permeability reservoirs. Candidates for acid fracturing are formations such as limestones ($CaCO_3$) or dolomites ($CaCO_3 \times MgCO_3$).

These materials react easily with hydrochloric acid to form chlorides and carbon dioxide. In comparison with the fracturing technique with proppants, acid fracturing has the advantage that no problem with proppant cleanout will appear. The acid etches the fracture faces unevenly, which on closure retain a highly conductive channel for the reservoir fluid to flow into the wellbore (Mukherjee and Cudney, 1992).

On the other hand, the length of the fracture is shorter, because the acid reacts with the formation and therefore is spent. If traces of fluoride are in the hydrochloric acid, then insoluble calcium fluoride is precipitated out. Therefore, plugging by the precipitate can jeopardize the desired effect of stimulation.

Encapsulated Acids

Acids, in particular, and etching agents, in general, may be mixed with a gelling agent and encapsulated with oils and polymers (Gonzalez and Looney (2000, 2001)).

In situ Formation of Acids

While acid fracturing is usually applicable for carbonate formations, acid fracturing in sandstone formations has been not a common practice due to the low rock solubility of mud acid. Methods and compositions that are useful in effectively acid fracturing of sandstone formations have been developed. Such a method of acid fracturing consists of (Qu and Wang, 2010):

1. Injecting into the formation an acid fracturing fluid at a pressure sufficient to form fractures within the formation, the acid fracturing fluid comprising a sulfonate ester, a fluoride salt, a proppant, and water, the sulfonate ester being hydrolyzed to produce sulfonic acid.
2. Producing hydrofluoric acid in situ in the formation by reacting the sulfonic acid with the fluoride salt subsequent to injection of the acid fracturing fluid into the formation.

Generating hydrofluoric acid in situ makes it possible to perform acid fracturing of sandstone formations with the assistance of partial monolayers of effectively placed propping agents and to create enlarged propped fractures. Increased fracture width will lead to higher fracture conductivity and enhanced hydrocarbon production than what is generally achieved from a conventional propped fracturing treatment.

Fluid Loss

Fluid loss limits the effectiveness of acid fracturing treatments. Therefore, formulations to control the fluid loss have been developed and characterized (Sanford et al., 1992; White et al., 1992). It was discovered that viscosifying the acid showed a remarkable improvement in acid fluid loss control. The enhancement was most pronounced in very low-permeability limestone cores. The nature of the viscosifying agent also influenced the success. Polymeric materials were more effective than surfactant-type viscosifiers (Gdanski, 1993).

A viscosity controlled acid contains gels that break back to the original viscosity 1 day after being pumped. These acids have been used both for matrix acidizing and for fracture acidizing to obtain longer fractures. The pH of the fluid controls the gel formation and breaking. The gels are limited to formation temperatures of 50–135 °C (Yeager and Shuchart, 1997).

Gel Breaker for Acid Fracturing

A particulate gel breaker for acid fracturing for gels crosslinked with titanium or zirconium compounds is composed of complexing materials such as fluoride, phosphate, sulfate anions, and multicarboxylated compounds. The particles are coated with a water-insoluble resin coating, which reduces the rate of release of the breaker materials of the particles so that the viscosity of the gel is reduced at a retarded rate (Boles et al., 1996).

SPECIAL PROBLEMS

Corrosion Inhibitors

Water-soluble 1,2-dithiol-3-thiones for fracturing fluids and other workover fluids have been described as corrosion inhibitors for aqueous environments (Oude Alink, 1993). These compounds are prepared by reacting a poly(ethylene oxide) that is capped with isopropylphenol with elemental sulfur.

The compounds perform better in aqueous systems than their nonoxylated analogs. The concentration range is usually in the 10–500 ppm range, based on the weight of the water in the system.

Iron Control in Fracturing

Results from laboratory tests and field jobs show that iron presents a significant and complex problem in stimulation operations (Smolarchuk and Dill, 1986). The problem presented by an acidizing fluid differs from that presented by a nonacidic or weakly acidic fracturing fluid. In general, acid dissolves iron compounds from the equipment and the flow lines as it is mixed and pumped into the formation.

The acid may dissolve additional iron as it reacts with the formation. If the fluid does not contain an effective iron control system, the dissolved iron could precipitate. This precipitate may then accumulate as it is carried toward the wellbore during flowback. This accumulation of solids may decrease the natural and the created permeability and can have a detrimental effect on the recovery of the treating fluid and production.

Iron can be controlled with certain complexing agents, in particular glucono-δ-lactone, citric acid, ethylenediamine tetraacetic acid, nitrilotriacetic acid, hydroxyethylethylenediaminetriacetic acid, hydroxyethyliminodiacetic acid, and the salts from those compounds. These compounds must be added together with nitrogen-containing compounds such as hydroxylamine salts or hydrazine salts (Dill et al., 1988; Frenier, 2001; Walker et al., 1987). Some complexing agents are shown in Figure 2.2.

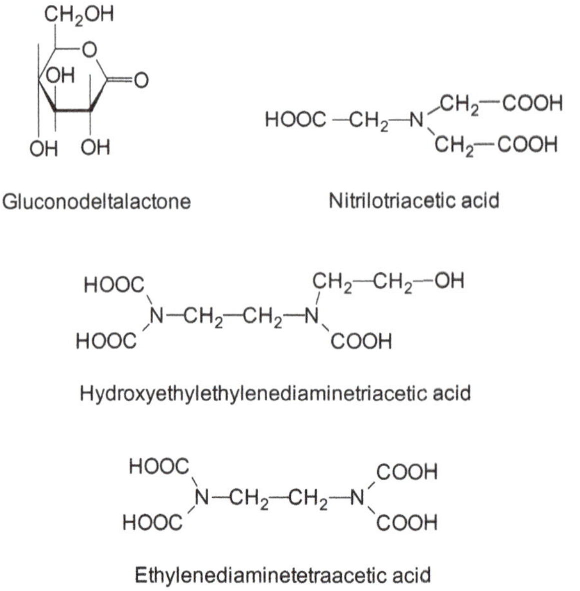

Gluconodeltalactone

Nitrilotriacetic acid

Hydroxyethylethylenediaminetriacetic acid

Ethylenediaminetetraacetic acid

FIGURE 2.2 Complexing agents for iron control.

In general, chelating agents possess some unique chemical characteristics. The most significant attribute of these chemicals is the high solubility of the free acids in aqueous solutions. Linear core flood tests were used to study the formation of wormholes.

Both hydroxyethylethylenediaminetriacetic acid and hydroxyethyliminodiacetic acid produced wormholes in limestone cores when tested at 65 °C (150 °F). However, the efficiency and capacities differ. Because these chemicals have a high solubility in the acidic pH range, it was possible to test acidic formulations with pH of <3.5 (Frenier et al., 2001).

To control the iron in an aqueous fracturing fluid having a pH below 7.5, a thioalkyl acid may be added (Brezinski et al., 1994). This is a reducing agent for the ferric ion, contrary to the complexants described in the previous paragraph.

Enhanced Temperature Stability

During the initial fracturing process, a degradation, which results in a decrease of viscosity, is undesirable. The polymer in fracturing fluids will degrade at elevated temperatures.

One method to prevent degradation too early is to cool down the formation with large volumes of pad solution before the fracturing job. Furthermore, the temperature stability of the fracturing fluid is extended through the addition of quantities of unhydrated, particulate guar or guar-derivative polymers before pumping the fracturing fluid into the formation (Nimerick and Boney, 1992).

Finally, the adjustment of the pH to moderate alkaline conditions can improve the stability.

The temperature resistance of guar gum can be improved by silanization (Zhang and Chen, 2012). The optimal reaction conditions are a reaction temperature of 85 °C and a molar ratio of guar gum to trimethylsilane chloride of 5:1. The viscosity of a silanized guar gum-based aqueous gel was greatly improved even at a high temperature of 80 °C.

The preferred crosslinkers for high-temperature applications are zirconium compounds. A formulation for a high-temperature guar-based fracturing fluid is given in Table 2.3. The fracturing fluid exhibits good viscosity and is stable at moderate to high temperatures of 80–120 °C.

Chemical Blowing

The efficiency of a fracturing fluid produced back from a formation can be increased by adding blowing agents (Abou-Sayed and Hazlett, 1989; Jennings, 1995). After placing the blowing agent, for example, agglomerated particles and granules containing the blowing agent and fracturing the formation, the blowing agent decomposes. Blowing agents are summarized in Table 2.4. Thereby the filter cake too becomes more porous or provides a driving force for the removal of fluid load from the matrix.

TABLE 2.3 Formulation for a High-Temperature Guar-Based Fracturing Fluid (Brannon et al., 1989)

Component	Action
Guar gum	Thickener
Zirconium or hafnium compound	Crosslinking agent
Bicarbonate salt	Buffer

TABLE 2.4 Blowing Agents (Jennings, 1995)

Chemical Compound
Dinitrosopentamethylenetetramine
Sodium hydrogen carbonate
p-Toluene sulfonyl hydrazide
Azodicarbonamide
p,*p*′-Oxybis(benzenesulfonyl hydrazide)

TABLE 2.5 Frost-Resistant Formulation for Hydraulic Fracturing Fluids (Barsukov et al., 1993)

Component	(%)
Hydrocarbon phase[a]	2–20
Surfactant	
Mineralized water	
Sludge from production of sulfonate additives (hydrocarbons 10–30%, calcium sulfonate 20–30%, calcium carbonate, and hydroxide 18–40%)[b]	10–35
Emultal[c]	0.5–2.0

[a]*Gas condensate, oil, or benzene.*
[b]*Slows down filtration and increases sand-holding capability, frost resistance, and stability.*
[c]*Surfactant-emulsifier.*

Increased porosity enhances the communication between the formation and the fracture, thus increasing the efficiency of the production of the fracturing fluid. The gas liberation within the matrix establishes the communication pathways for subsequent fracture and the well.

Frost-Resistant Formulation

A frost-resistant formulation is given in Table 2.5. The composition has a frost resistance of −35 to −45 °C (Barsukov et al., 1993).

Formation Damage in Gas Wells

Studies (Gall et al., 1988) on formation damage using artificially fractured, low-permeability sandstone cores indicated that viscosified fracturing fluids can severely restrict the gas flow through narrow fractures. Poly(saccharide) polymers such as hydroxypropyl guar, HEC, and xanthan gum caused a significant reduction of the gas flow through the cracked cores, up to 95%.

In contrast, PAM gels caused little or no reduction in the gas flow through cracked cores after a liquid cleanup. Other components of fracturing fluids, such as surfactants and breakers, caused less damage to gas flows.

CHARACTERIZATION OF FRACTURING FLUIDS

Historically, viscosity measurements have been the single most important method to characterize fluids in petroleum-producing applications. Whereas the ability to measure a fluid's resistance to flow has been available in the laboratory for a long time, a need to measure the fluid properties at the well site has prompted the development of more portable and less sophisticated viscosity-measuring devices (Parks et al., 1986).

These instruments must be durable and simple enough to be used by persons with a wide range of technical skills. As a result, the Marsh funnel and the Fann concentric cylinder, both variable-speed viscometers, have found wide use. In some instances, the Brookfield viscometer has also been used.

However, it has been established that an intense control of certain variables may improve the execution of a hydraulic fracturing job and the success of a stimulation. Therefore, an intense quality control is recommended (Ely, 1996; Ely et al., 1990). Such a program includes monitoring the breaker performance at low temperatures and measuring the sensitivity of fracturing fluids to variations in crosslinker loading, temperature stabilizers, and other additives at higher temperatures.

Rheologic Characterization

To design a successful hydraulic fracturing treatment employing crosslinked gels, accurate measurements of the rheologic properties of these fluids are required. Rheologic characterization of borate crosslinked gels turned out to be difficult with a rotational viscometer. In a laboratory apparatus, field pumping conditions (i.e., crosslinking the fluid on the fly) and fluid flow down tubing or casing and in the fracture could be simulated (Shah et al., 1988). The effects of the pH and temperature of the fluid and the type and concentration of the gelling agent on the rheologic properties of fluids have been measured.

These parameters have significant effect on the final viscosity of the gel in the fracture. Correlations to estimate friction pressures in field size tubulars have been developed from laboratory test data. In conjunction with field calibrations, these correlations can aid in accurate prediction of the friction pressure of borate crosslinked fluids.

Zirconium-Based Crosslinking Agent

The concentration of a crosslinking agent containing zirconium in a gel is determined by first adding an acid to break the gel and converting the zirconium into the ionic noncomplexed form (Chakrabarti and Marczewski, 1990). This is followed by the addition of Arsenazo (III) to produce a colored complex, which can be determined with standard colorimetric methods. Arsenic compounds are highly toxic. The colorimetric reagent to measure zirconium in gels is shown in Figure 2.3.

Oxidative Gel Breaker

The concentration of an oxidative gel breaker can be measured by colorimetric methods, by periodically or continuously sampling the gel (Chakrabarti et al.,

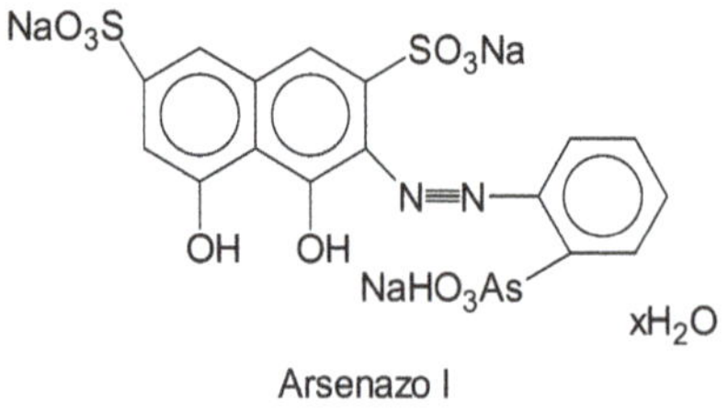

Arsenazo I

FIGURE 2.3 Colorimetric reagent to measure zirconium in gels.

1988). The colorimetric reagent is sensitive to oxidizing agents. It contains iron ions and thiocyanate. Thus the quantity of breaker added to the fracturing fluid can be controlled. The method is based on the oxidation of ferrous ions to ferric ions

$$Fe^{2+} \rightarrow Fe^{3+} + e^{-},$$

which form a deep red complex with thiocyanate.

Size Exclusion Chromatography

Size exclusion chromatography (Brannon and Tjon, 1995; Gall and Raible, 1986) has been used to monitor the degradation of the thickeners initiated by various oxidative and enzymatic breakers.

The research revealed that a partially broken or unbroken polymer may result in a significant reduction of the flow through a porous medium. An insoluble residue was generated during the degradation of guar polymers. This residue can affect the pore size of the medium (Kyaw et al., 2012).

Assessment of Proppants

There are standardized methods for the characterization of the effect of proppants (ISO-13503-5, 2006; API Standard RP 19C, 2008). The general methods of assessment have been exemplified (Wen et al., 2007).

For some proppants, the experiments reveal relationships between the long-term fracture conductivity and the closure pressure as a polynomial (Wen et al., 2007)

$$F = A_1 + A_2 P + A_3 P^2 + A_4 P^3, \tag{2.1}$$

where F is the long-term fracture conductivity, P is the closure pressure, and A_i are some constants. Similarly, an exponential relationship between fracture conductivity F and time t has be obtained for a certain closure pressure (Wen et al., 2007),

$$F = \exp(-A_5 t) + A_6. \tag{2.2}$$

Tradenames in References

Tradename	Supplier
Description	
WS-44	Halliburton Energy Services, Inc.
Emulsifier (Welton et al., 2010)	

REFERENCES

Abou-Sayed, I.S., Hazlett, R.D., 1989. Removing fracture fluid via chemical blowing agents. US Patent 4 832 123, assigned to Mobil Oil Corp., 23 May 1989.

Allan, T.L., Amin, J., Olson, A.K., Pierce, R.G., 2008. Fracturing fluid containing amphoteric glycinate surfactant. US Patent 7 399 732, assigned to Calfrac Well Services Ltd., Calgary, Alberta, CA, Chemergy Ltd., Calgary, Alberta, CA, 15 July 2008. <http://www.freepatentsonline.com/7399732.html>.

Anonymous, 1999. Fracturing products and additives. World Oil 220 (8), 135,137,139–145.

API Standard RP 19C, 2008. Recommended practice for measuring the long-term conductivity of proppants. API Standard API RP 19C, American Petroleum Institute, Washington, DC.

Barsukov, K.A., Ismikhanov, V.Y., Akhmetov, A.A., Pop, G.S., Lanchakov, G.A., Sidorenko, V.M., 1993. Composition for hydro-bursting of oil and gas strata—consists of hydrocarbon phase, sludge from production of sulphonate additives to lubricating oils, surfactant-emulsifier and mineralised water. SU Patent 1 794 082, assigned to Urengoi Prod. Assoc., 7 February 1993.

Boles, J.L., Metcalf, A.S., Dawson, J.C., 1996. Coated breaker for crosslinked acid. US Patent 5 497 830, assigned to BJ Services Co., 12 March 1996.

Bonekamp, J.E., Rose, G.D., Schmidt, D.L., Teot, A.S., Watkins, E.K., 1993. Viscoelastic surfactant based foam fluids. US Patent 5 258 137, assigned to Dow Chemical Co., 2 November 1993.

Brannon, H.D., Tjon, J.P.R.M., 1995. Characterization of breaker efficiency based upon size distribution of polymeric fragments. In: Proceedings Volume Annual SPE Technical Conference, Dallas, 22–25 October 1995, pp. 415–429.

Brannon, H.D., Hodge, R.M., England, K.W., 1989. High temperature guar-based fracturing fluid. US Patent 4 801 389, assigned to Dowell Schlumberger Inc., 31 January 1989.

Brezinski, M., Gardner, T.R., Harms, W.M., Lane Jr., J.L., King, K.L., 1994. Controlling iron in aqueous well fracturing fluids. EP Patent 599 474, assigned to Halliburton Co., 1 June 1994.

Chakrabarti, S., Marczewski, C.Z., 1990. Determining the concentration of a cross-linking agent containing zirconium. GB Patent 2 228 996, assigned to British Petroleum Co. Ltd., 12 September 1990.

Chakrabarti, S., Martins, J.P., Mealor, D., 1988. Method for controlling the viscosity of a fluid. GB Patent 2 199 408, assigned to British Petroleum Co. Ltd., 6 July 1988.

Chatterji, J., King, B.J., King, K.L., 2007. Recyclable foamed fracturing fluids and methods of using the same. US Patent 7 205 263, assigned to Halliburton energy Services, Inc., Duncan, OK, 17 April 2007. <http://www.freepatentsonline.com/7205263.html>.

Cottrell, T.L., Spronz, W.D., Weeks III, W.C., 1988. Hugoton infill program uses optimum stimulation technique. Oil Gas J. 86 (28), 88–90

Deng, Q., Xu, J., Gu, X., Tang, Y., 2012. Properties evaluation of polymer-free fluid for fracturing application. Adv. Mater. Res. 482–484 (Pt. 2, Advanced Composite Materials), 1180–1183. http://dx.doi.org/10.4028/www.scientific.net/AMR.482-484.1180.

Dill, W.R., Ford, W.G.F., Walker, M.L., Gdanski, R.D., 1988. Treatment of iron-containing subterranean formations. EP Patent 258 968, 9 March 1988.

Ely, J.W., 1996. How intense quality control improves hydraulic fracturing. World Oil 217 (11), 59–60, 62–65, 68.

Ely, J.W., Wolters, B.C., Holditch, S.A., 1990. Improved job execution and stimulation success using intense quality control. In: Proceedings Volume 37th Annual Southwestern Petroleum Short Course Association et al Meeting, Lubbock, Texas, 18–19 April 1890, pp. 101–114.

Frenier, W.W., 2001. Well treatment fluids comprising chelating agents. WO Patent 0 183 639, assigned to Sofitech NV, Schlumberger Serv. Petrol, Schlumberger Canada Ltd., Schlumberger Technol. BV, and Schlumberger Holdings Ltd., 8 November 2001.

Frenier, W.W., Fredd, C.N., Chang, F., 2001. Hydroxyaminocarboxylic acids produce superior formulations for matrix stimulation of carbonates. In: Proceedings Volume SPE Europe Formation Damage Conference, the Hague, Netherlands, 21–22 May 2001).

Gall, B.L., Raible, C.J., 1986. The use of size exclusion chromatography to study the degradation of water-soluble polymers used in hydraulic fracturing fluids. In: Proceedings 192nd ACS National Meetings, vol. 55. American Chemical Society Polymeric Materials: Science and Engineering Division Technology Program, Anaheim, Calif, 7–12 September 1986, pp. 572–576.

Gall, B.L., Maloney, D.R., Raible, C.J., Sattler, A.R., 1988. Permeability damage to natural fractures caused by fracturing fluid polymers. In: Proceedings Volume SPE Rocky Mountain Regional Meeting, Casper, Wyo, 11–13 May 1988, pp. 551–560.

Gdanski, R.D., 1993. Fluid properties and particle size requirements for effective acid fluid-loss control. In: Proceedings Volume SPE Rocky Mountain Regional Meeting: Low Permeability Reservoirs Symposium, Denver, 26–28 April 1993, pp. 81–94.

Gonzalez, M.E., Looney, M.D., 2000. The use of encapsulated acid in acid fracturing treatments. WO Patent 0 075 486, assigned to Texaco Development Corp., 14 December 2000.

Gonzalez, M.E., Looney, M.D., 2001. Use of encapsulated acid in acid fracturing treatments. US Patent 6 207 620, assigned to Texaco Inc., 27 March 2001.

Harris, P.C., 1988. Fracturing-fluid additives. J. Pet. Technol. 40 (10), 1277–1279.

Holditch, S.A., Xiong, H., Rahim, Z., Rueda, J., 1993. Using an expert system to select the optimal fracturing fluid and treatment volume. In: Proceedings Volume SPE Gas Technology Symposium, Calgary, Canada, 28–30 June 1993, pp. 515–527.

Hrachovy, M.J., 1994a. Hydraulic fracturing technique employing in situ precipitation. WO Patent 9 406 998, assigned to Union Oil Co. California, 31 March 1994.

Hrachovy, M.J., 1994b. Hydraulic fracturing technique employing in situ precipitation. US Patent 5 322 121, assigned to Union Oil Co. California, 21 June 1994.

ISO-13503-5, 2006. Petroleum and natural gas industries—completion fluids and materials—Part 5: Procedures for measuring the long-term conductivity of proppants. ISO Standard ISO-13503-5, International Organization for Standardization, Geneva, Switzerland.

Jennings, Jr., A.R., 1995. Method of enhancing stimulation load fluid recovery. US Patent 5 411 093, assigned to Mobil Oil Corp., 2 May 1995.

Kyaw, A., Nor Azahar, B.S., Tunio, S.Q., 2012. Fracturing fluid (guar polymer gel) degradation study by using oxidative and enzyme breaker. Res. J. Appl. Sci. Eng. Technol. 4 (12), 1667–1671. <http://maxwellsci.com/print/rjaset/v4-1667-1671.pdf>.

Lukocs, B., Mesher, S., Wilson, T.P.J., Garza, T., Mueller, W., Zamora, F., Gatlin, L.W., 2007. Non-volatile phosphorus hydrocarbon gelling agent. US Patent Application 20070173413, assigned to Clearwater International, LLC, 26 July 2007. <http://www.freepatentsonline.com/20070173413.html>.

Middaugh, R.L., Harris, P.C., Heath, S.J., Taylor, R.S., Hoch, O.F., Phillippi, M.L., Slabaugh, B.F., Terracina, J.M., 2007. Coarse-foamed fracturing fluids and associated methods. US Patent 7 261 158, assigned to Halliburton Energy Services, Inc., Duncan, OK, 28 August 2007. <http://www.freepatentsonline.com/7261158.html>.

Mukherjee, H., Cudney, G., 1992. Extension of acid fracture penetration by drastic fluid-loss control. SPE Unsolicited Pap.

Nimerick, K.H., Boney, C.L., 1992. Method of fracturing high temperature wells and fracturing fluid therefore. US Patent 5 103 913, assigned to Dowell Schlumberger Inc., 14 April 1992.

Oude Alink, B.A., 1993. Water soluble 1,2-dithio-3-thiones. US Patent 5 252 289, assigned to Petrolite Corp., 12 October 1993.

Parks, C.F., Clark, P.E., Barkat, O., Halvaci, J., 1986. Characterizing polymer solutions by viscosity and functional testing. In: Proceedings of the 192nd ACS National: Meeting, vol. 55. American Chemical Society Polymeric Materials: Science and Engineering Division Technology Program, Anaheim, Calif, 7–12 September 1986, pp. 880–888.

Penny, G.S., Conway, M.W., 1989. Fluid Leakoff. Recent Advances In Hydraulic Fracturing (SPE Henry L. Doherty Monogr Ser), vol. 12. SPE, Richardson, Texas, pp. 147–176.

Qu, Q., Wang, X., 2010. Method of acid fracturing a sandstone formation. US Patent 7 704 927, assigned to BJ Services Co., Houston, TX, 27 April 2010. <http://www.freepatentsonline.com/7704927.html>.

Sanford, B.D., Dacar, C.R., Sears, S.M., 1992. Acid fracturing with new fluid-loss control mechanisms increases production, little knife field, North Dakota. In: Proceedings Volume SPE Rocky Mountain Regional Meeting, Casper, Wyo, 18–21 May 92, pp. 317–324.

Shah, S.N., Harris, P.C., Tan, H.C., 1988. Rheological characterization of borate crosslinked fracturing fluids employing a simulated field procedure. In: Proceedings Volume SPE Production Technology Symposium, Hobbs, New Mexico, 7–8 November 1988.

Smolarchuk, P., Dill, W., 1986. Iron control in fracturing and acidizing operations. In: Proceedings Volume 37th Annual CIM Petroleum Society Technical Meeting, Calgary, Canada, vol. 1, 8–11 June 1986, pp. 391–397.

Stacy, A.L., Weber, R.B., 1995. Method for reducing deleterious environmental impact of subterranean fracturng processes. US Patent 5 424 285, assigned to Western Co. North America, 13 June 1995.

Tamayo, H.C., Lee, K.J., Taylor, R.S., 2008. Enhanced aqueous fracturing fluid recovery from tight gas formations: Foamed CO_2 pre-pad fracturing fluid and more effective surfactant systems. J. Can. Pet. Technol. 47 (10), 33–38. http://dx.doi.org/10.2118/08-10-33.

Walker, M.L., Ford, W.G.F., Dill, W.R., Gdanski, R.D., 1987. Composition and method of stimulating subterranean formations. US Patent 4 683 954, 4 August 1987.

Wang, Z., Wang, S., Sun, X., 2012. The influence of surfactant concentration on rheology and proppant-carrying capacity of VES fluids. Adv. Mater. Res. 361–363 (Pt. 1, Natural Resources and Sustainable Development), 574–578. http://dx.doi.org/10.4028/www.scientific.net/AMR.361-363.574.

Welton, T.D., Todd, B.L., McMechan, D., 2010. Methods for effecting controlled break in pH dependent foamed fracturing fluid. US Patent 7 662 756, assigned to Halliburton Energy Services, Inc., Duncan, OK, 16 February 2010. <http://www.freepatentsonline.com/7662756.html>.

Wen, Q., Zhang, S., Wang, L., Liu, Y., Li, X., 2007. The effect of proppant embedment upon the long-term conductivity of fractures. J. Pet. Sci. Eng. 55 (3–4), 221–227. <http://www.sciencedirect.com/science/article/B6VDW-4M6459G-2/2/977b6f4fd9756baba71ba40771815d40>.

White, D.J., Holms, B.A., Hoover, R.S., 1992. Using a unique acid-fracturing fluid to control fluid loss improves stimulation results in carbonate formations. In: Proceedings Volume SPE Permian Basin Oil and Gas Recovery Conference, Midland, Texas, 18–20 March 1992, pp. 601–610.

Xiong, H., Davidson, B., Saunders, B., Holditch, S.A., 1996. A comprehensive approach to select fracturing fluids and additives for fracture treatments. In: Proceedings Volume Annual SPE Technical Conference, Denver, 6–9 October 1996, pp. 293–301.

Yeager, V., Shuchart, C., 1997. In situ gels improve formation acidizing. Oil Gas J. 95 (3), 70–72.

Zhang, J., Chen, G., 2012. Improvement of the temperature resistance of guar gum by silanization. Adv. Mater. Res. 415–417 (Pt. 1, Advanced Materials), 652–655. http://dx.doi.org/10.4028/www.scientific.net/AMR.415-417.652.

Chapter 3

Thickeners

A variety of compounds useful as thickeners are shown in Table 3.1. Subsequently, the individual compounds are explained in detail.

Polymers

Thickener polymers include poly(urethane)s, poly(ester)s, and poly (acrylamide)s, further, natural polymers, and modified natural polymers (Doolan and Cody, 1995).

pH Responsive Thickeners

The viscosity of ionic polymers is dependent on the pH. In particular, pH responsive thickeners can be prepared by copolymerization of acrylic or methacrylic acid ethyl acrylate or other vinyl monomers and tristyrylpoly(ethylene-oxy)$_x$ methyl acrylate. Such a copolymer provides a stable aqueous colloidal dispersion at an acid pH lower than 5.0 but becomes an effective thickener for aqueous systems upon adjustment to a pH of 5.5–10.5 or higher (Robinson (1996,1999)).

Mixed Metal Hydroxides

By addition of mixed metal hydroxides, typical bentonite muds are transformed to an extremely shear thinning fluid (Lange and Plank, 1999). At rest these fluids exhibit a very high viscosity but are thinned to an almost water-like consistency when shear stress is applied.

In theory, the shear thinning rheology of mixed metal hydroxides and bentonite fluids is explained by the formation of a three-dimensional, fragile network of mixed metal hydroxides and bentonite.

The positively charged mixed metal hydroxide particles attach themselves to the surface of negatively charged bentonite platelets. Typically, magnesium aluminum hydroxide salts are used as mixed metal hydroxides.

Mixed metal hydroxides demonstrate the following advantages in drilling (Felixberger, 1996):

Hydraulic Fracturing Chemicals and Fluids Technology. http://dx.doi.org/10.1016/B978-0-12-411491-3.00003-0
© 2013 Elsevier Inc. All rights reserved.

TABLE 3.1 Thickeners

Compound	References
A water-soluble copolymer of hydrophilic and hydrophobic monomers, acrylamide (AAm)-acrylate of silane or siloxane	Meyer et al. (1999)
Carboxymethyl cellulose, poly(ethylene glycol)	Lundan et al. (1993) and Lundan and Lahteenmaki (1996)
Combination of a cellulose ether with clay	Rangus et al. (1993)
Amide-modified carboxyl-containing poly(saccharide)	Batelaan and van der Horts (1994)
Sodium aluminate and magnesium oxide	Patel (1994)
Thermally stable hydroxyethyl cellulose (HEC) 30% ammonium or sodium thiosulfate and 20% HEC	Lukach and Zapico (1994)
Acrylic acid (AA) copolymer and oxyalkylene with hydrophobic group	Egraz et al. (1994)
Copolymers acrylamide-acrylate and vinylsulfonate-vinylamide	Waehner (1990)
Cationic poly(galactomannan)s and anionic xanthan gum	Yeh (1995b)
Copolymer from vinyl urethanes and AA or alkyl acrylates	Wilkerson et al. (1995)
2-Nitroalkyl ether-modified starch	Gotlieb et al. (1996)
Polymer of glucuronic acid	Courtois-Sambourg et al. (1993)
Ferrochrome lignosulfonate and carboxymethyl cellulose	Kotelnikov et al. (1992)
Cellulose nanofibrils[a]	Langlois (1998) and Langlois et al. (1999)
Quaternary alkyl amido ammonium salts	Subramanian et al. (2001a)
Chitosan[b]	House and Cowan (2001)

[a] *Stable up to temperatures of about 180 °C.*
[b] *Solubilized in acidic solution.*

- High cuttings removal.
- Suspension of solids during shutdown.
- Lower pump resistance.
- Stabilization of the borehole.
- High drilling rates.
- Protection of the producing formation.

Mixed metal hydroxide drilling muds have been successfully used in horizontal wells, in tunneling under rivers, roads, and bays, for drilling in fluids, for drilling large-diameter holes, with coiled tubing, and to ream out cemented pipe.

Mixed metal hydroxides can be prepared from the corresponding chlorides treated with ammonium (Burba and Strother, 1991). Experiments done with various drilling fluids showed that the mixed metal hydroxides system, coupled with propylene glycol (Deem et al., 1991), caused the least skin damage of the drilling fluids tested.

Thermally activated mixed metal hydroxides, made from naturally occurring minerals, especially hydrotalcites, may contain small or trace amounts of metal impurities besides the magnesium and aluminum components, which are particularly useful for activation (Keilhofer and Plank, 2000).

Mixed hydroxides of bivalent and trivalent metals with a three-dimensional spaced-lattice structure of the garnet type ($Ca_3Al_2[OH]_{12}$) have been described (Burba et al., 1992; Mueller et al., 1997).

THICKENERS FOR WATER-BASED SYSTEMS

A gelling agent is also known as a viscosifying agent, and refers to a material that can make the fracturing fluid into a gel, thereby increasing its viscosity (Welton et al., 2010).

Suitable gelling agents include guar gum, xanthan gum, welan gum, locust bean gum, gum ghatti, gum karaya, tamarind gum, and tragacanth gum. Guar gum can be functionalized to hydroxyethyl guar, hydroxypropyl guar, or carboxymethyl guar. Examples of water-soluble cellulose ethers include methyl cellulose, carboxymethyl cellulose (CMC), HEC, and hydroxyethyl carboxymethyl cellulose (Welton et al., 2010).

Artificial polymers, such as copolymers from AAm, methacrylamide, AA, or methacrylic acid, or those from 2-acrylamido-2-methyl-1-propane sulfonic acid (AMPS) derivates and *N*-vinylpyridine, have all been used (Welton et al., 2010), but also naturally occurring polysaccharides and their derivatives (Lemanczyk, 1992). They can increase the viscosity of the fluid when used in comparatively small amounts. Table 3.2 summarizes polymers suitable for fracturing fluids.

Guar is shown in Figure 3.1. In hydroxypropyl guar, some of the hydroxyl groups are etherified with oxopropyl units. Compositions for gelling a hydrocarbon fracturing fluid are basically different from those for aqueous fluids. A possible formulation consists of a gelling agent, a phosphate ester, a crosslinking agent, a multivalent metal ion, and a catalyst, a fatty quaternized amine (Lawrence and Warrender, 2010).

Zirconium-based Crosslinking Composition

Some commercially available zirconate crosslinkers, such as tetra triethanol amine zirconate, crosslink too fast at high pH conditions, thus causing a significant loss in viscosity due to shear degradation (Putzig, 2012).

TABLE 3.2 Summary of Thickeners Suitable for Fracturing Fluids

Thickener	References
Hydroxypropyl guar[a]	
Galactomannans[b]	Mondshine (1987)
HEC-modified vinylphosphonic acid	Holtmyer and Hunt (1992)
Carboxymethyl cellulose	
Polymer from *N*-vinyl lactam monomers, vinylsulfonates[c]	Bharat (1990)
Reticulated bacterial cellulose[d]	Westland et al. (1993)
Bacterial xanthan[e]	Hodge (1997)

[a] *General-purpose eightfold power of thickening in comparison to starch.*
[b] *Increased temperature stability, used with boron-based crosslinkers.*
[c] *High-temperature stability.*
[d] *Superior fluid performance.*
[e] *Imparts high viscosity.*

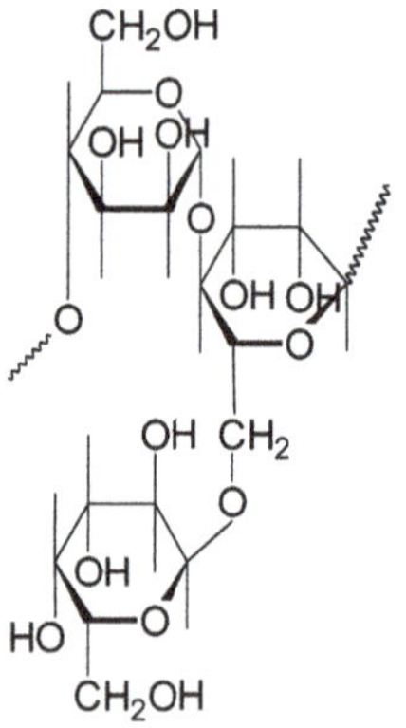

FIGURE 3.1 Structural unit of guar.

In contrast, other zirconium complexes of triethanol amine (Rummo and Startup, 1986; Kucera, 1987; Baranet et al., 1987) can be used, but suffer from a slow rate of crosslinking (Putzig, 2012). In addition, a similar loss in viscosity due to shear degradation is observed.

A water-based crosslinking composition has been developed that contains a pH buffer, crosslinkable organic polymer, and a solution of a zirconium crosslinking agent (Putzig, 2012). The zirconium complex is with an alkanolamine and ethylene glycol.

A most suitable is tetraalkyl zirconate is tetra-*n*-propyl zirconate, available as TYZOR NPZ®, a solution in *n*-propanol, with a zirconium content as ZrO_2 of about 28% by weight.

The composition further comprises a crosslinkable organic polymer, e.g., guar deviates. However, CMC are advantageous when the pH of the composition is less than 6.0 or higher than 9.0, or when the permeability of the formation is such that one wishes to keep the residual solids at a low level to prevent damage to the formation (Putzig, 2012).

Guar

Guar is a branched polysaccharide from the guar plant Cyamopsis tetragonolobus, which originated in India, and is now found in the southern United States. It has a molar mass of approximately 220 kDa, and it consists of mannose in the main chain and galactose in the side chain. The ratio of mannose to galactose is 2:1.

Polysaccharides having this structure are referred to as *heteromannans*, and in particular as *galactomannans*. Derivatives of guar are therefore sometimes called *galactomannans*.

Guar-based gelling agents, typically hydroxypropyl guar, are widely used to viscosify fracturing fluids because of their desirable rheological properties, economics, and ease of hydration. Nonacetylated xanthan is a variant of xanthan gum, which interacts synergistically with guar to give superior viscosity and particle transport at lower polymer concentrations.

Static leakoff experiments with borate crosslinked and zirconate-crosslinked hydroxypropyl guar fluids showed practically the same leakoff coefficients (Zeilinger et al., 1991). An investigation of their stress-sensitive properties showed that zirconate filter cakes have viscoelastic properties, but borate filter cakes are merely elastic. Non-crosslinked fluids show no filter cake type behavior for a large range of core permeabilities, but rather, a viscous flow that is dependent on characteristics of the porous medium.

The addition of glycols, such as ethylene glycol (EG), to aqueous fluids gelled with guar gum can increase the viscosity of the fluid and stabilize the fluid brines. Such fluids are more stable at high temperatures from 27–177 °C (80–350 °F). The formation damage is minimized after hydraulic fracturing operations, as less of the guar polymer can be used, but the same viscosity is achieved by the addition of a glycol (Kelly et al., 2007).

The crosslinker can be a borate, a titanate, or a zirconate. The stability of the gel is improved by the addition of sodium thiosulfate. The development of the viscosity at 93 °C (200 °F) of brine fluids with 2.4 kg m^{-3} guar and 5% KCl, with varying amounts of EG, is shown in Figure 3.2.

By using the sodium derivative of ethylenediamine tetraacetic acid as gel breaker in these compositions with EG, the decay of viscosity with time can be adjusted accordingly (Kelly et al., 2007).

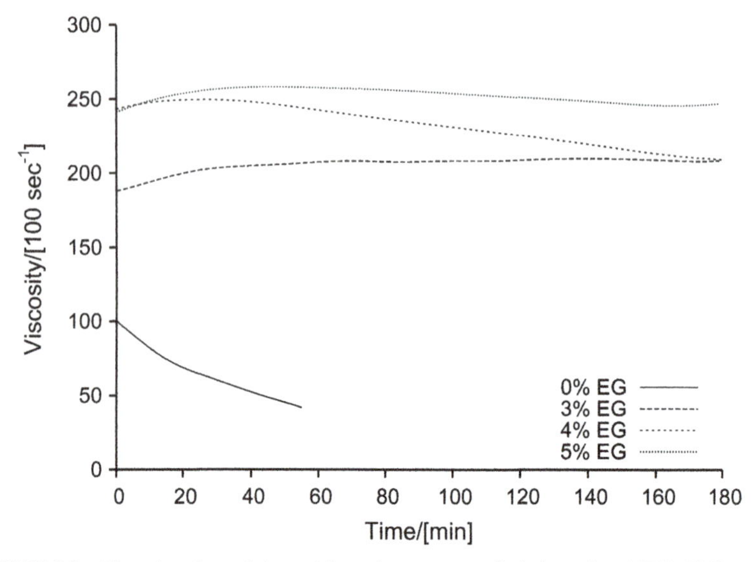

FIGURE 3.2 Viscosity of guar brines with varying amounts of ethylene glycol (EG) (Kelly et al., 2007).

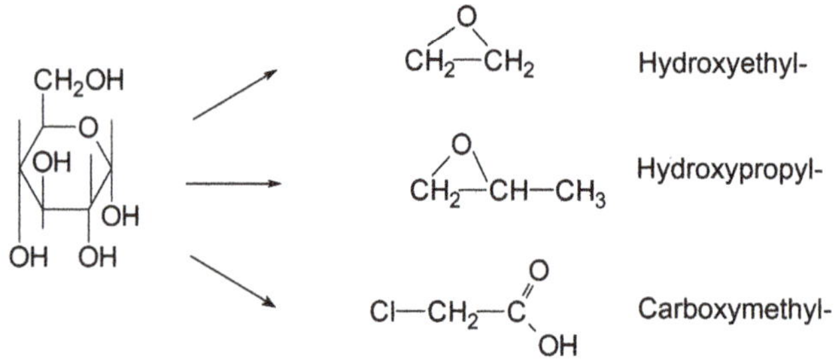

Hydroxyethyl-
Hydroxypropyl-
Carboxymethyl-

FIGURE 3.3 Modification of polyhydroxy compounds.

Anionic galactomannans, which are derived from guar gum by partially esterifying hydroxyl groups with sulfonate groups that result from AMPS and 1-allyloxy-2-hydroxypropyl sulfonic acid (Yeh, 1995a), have been claimed to be suitable as thickeners. The composition is capable of producing enhanced viscosities, when used alone, or in combination with a cationic polymer and distributed in a solvent.

Polyhydroxy compounds can be modified by various reactions. Etherification, exemplified with dextrose as the model compound, is shown in Figure 3.3. Vinyl compounds used for the modification of guar are shown in Figure 3.4.

The temperature stability of fracturing fluids containing galactomannan polymers is increased by adding a sparingly soluble borate with a slow solubility rate. This provides a source of boron for solubilizing at elevated temperatures, thus enhancing the crosslinking of the galactomannan polymer.

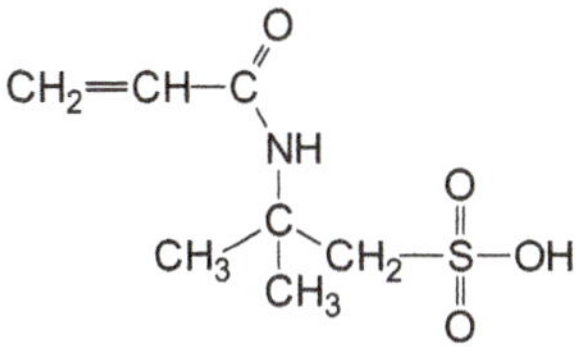

2-Acrylamido-2-methyl-1-propane sulfonic acid

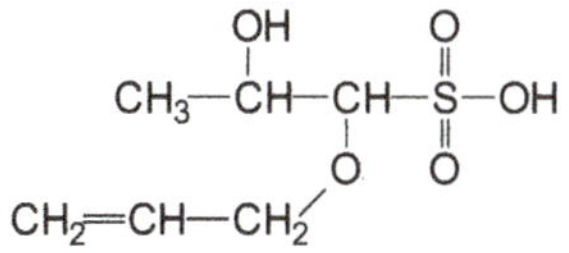

1-Allyloxy-2-hydroxypropyl sulfonic acid

FIGURE 3.4 Vinyl modifiers for guar gum.

The effect of nanoparticles on the rheology of fracturing fluids has been studied (Jafry et al., 2011). Vapor grown carbon fibers have been coated with silica and subsequently functionalized with octadecyltrichlorosilane. The so-modified fibers have been added to a fracturing fluid gel. Rheological measurements at pH of 8.6, 9.3, and 10.3, respectively, showed an interaction with the gels only at a lower pH than typically used for fracturing fluids.

The presence of the modified nanoparticles minimally increases the storage modulus of the guar gels. These gels behave similarly to the plain guar gels, displaying no permanent damage to the gels under the applied shear. In addition, the presence of the modified nanoparticles does not seem to alter the structure and the crosslinking sites of the guar gel (Jafry et al., 2011).

The grafting of a poly(alkoxyalkyleneamide) to guar gum gives suitable guar derivates. The modification of either guar gum or hydroxypropyl guar is achieved in a three-step process (Bahamdan and Daly, 2007):

1. Carboxymethylation with sodium chloroacetate.
2. Esterification with dimethyl sulfate.
3. Amidation with a poly(alkoxyalkyleneamide).

The process was followed by infrared spectroscopy. A series of hydroxypropyl guar derivates with degrees of carboxymethylation of 0.2–0.3 were modified with poly(alkoxyalkyleneamide) with molecule weights ranging from 300 to 3000 D.

The ratio of oxypropylene to oxyethylene units in the poly (alkoxyalkyleneamide) was varied from 9:1 to 8:58 in order to adjust the hydrophobicity of the final materials. The viscosities of the grafts are one to two orders of magnitude lower than the neat guar gum (Bahamdan and Daly, 2007).

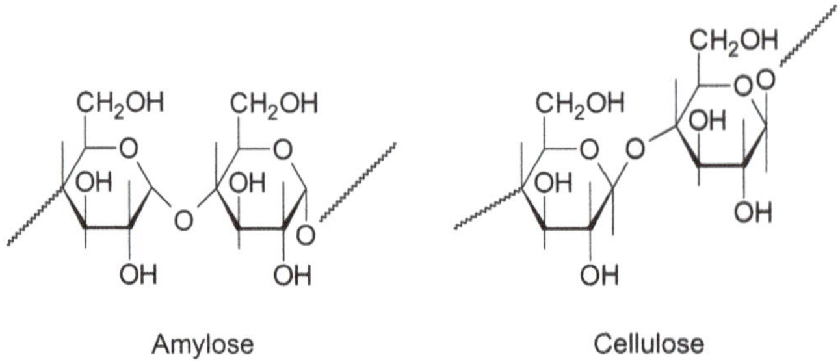

Amylose Cellulose

FIGURE 3.5 Amylose and cellulose.

Hydroxyethyl Cellulose

HEC can be chemically modified by the reaction with vinylphosphonic acid in the presence of the reaction product of hydrogen peroxide and a ferrous salt. The HEC forms a graft copolymer with the vinylphosphonic acid.

Amylose and cellulose are shown in Figure 3.5. Amylose is a linear polymer of glucose and is water soluble. The difference between amylose and cellulose is the way in which the glucose units are linked; amylose has α-linkages, whereas cellulose contains β-linkages. Because of this difference, amylose is soluble in water and cellulose is not. Chemical modification allows cellulose to become water soluble.

Modified HEC has been proposed as a thickener for hydraulic fracturing fluids (Holtmyer and Hunt, 1992). Polyvalent metal cations may be employed to crosslink the polymer molecules to further increase the viscosity of the aqueous fluid.

Biotechnologic Products

Gellan Gum and Wellan Gum

Gellan gum is the generic name of an extracellular polysaccharide produced by the bacterium *Pseudomonas elodea*. It is a linear anionic polysaccharide with a molecular mass of 500 kDa, consisting of 1,3-β-*D*-glucose, 1,4-β-*D*-glucuronic acid, 1,4-β-*D*-glucose, and 1,4-α-*L*-rhamnose.

Wellan gum is produced by aerobic fermentation. The backbone of wellan gum is identical to gellan gum, but it has a side chain consisting of *L*-mannose or *L*-rhamnose. It is used in fluid loss additives and is extremely compatible with calcium ions in alkaline solutions.

Reticulated Bacterial Cellulose

A cellulose produced from bacteria, with an intertwined reticulated structure, has unique properties and functionalities unlike conventional cellulose. It improves fluid rheology and particle suspension over a wide range of conditions in aqueous systems (Westland et al., 1993).

TABLE 3.3 Variant Xanthan Gums

Number	Repeating units	Ratio
Pentamer	*D*-glucose: *D*-mannose: *D*-glucuronic acid	2:2:1
Tetramer	*D*-glucose: *D*-mannose: *D*-glucuronic acid	2:1:1

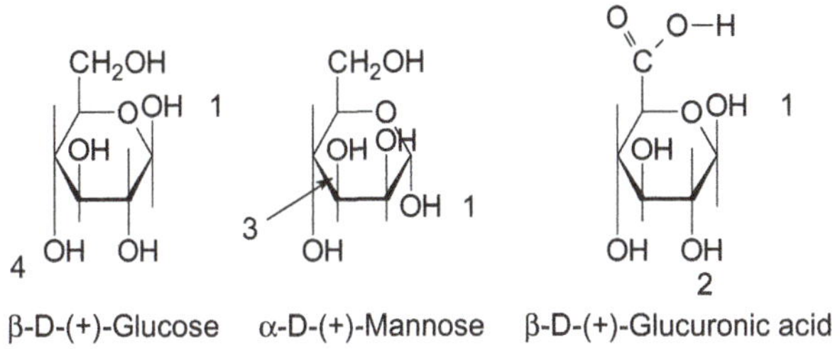

FIGURE 3.6 Carbohydrates and derivates.

Xanthan Gum

Xanthan gum is produced by the bacterium *Xanthomonas campestris* and has been used commercially since 1964. Xanthans are water-soluble polysaccharide polymers with the repeating units (Doherty et al., 1992) shown in Table 3.3 and Figure 3.6.

The *D*-glucose moieties are linked in a β-(1,4) configuration, and the inner *D*-mannose moieties are linked in an α-(1,3) configuration, generally alternating with glucose moieties. The *D*-glucuronic acid moieties are linked in a β-(1,2) configuration to the inner mannose moieties. The outer mannose moieties are linked to the glucuronic acid moieties in a β-(1,4) configuration.

Xanthan gum is used in oilfield applications in the form of a fermentation broth containing 8–15% of the polymer. The viscosity is less dependent on the temperature than for other polysaccharides.

Viscoelastic Formulations

Viscoelastic surfactant (VES) fluids have the following advantages over conventional polymer formulations (Li et al., 2010):

- Higher permeability in the oil-bearing zone.
- Lower formation or subterranean damage.
- Higher viscosifier recovery after fracturing.
- No need for enzymes or oxidizers to break down viscosity.
- Easier hydration and faster buildup to optimum viscosity.

Disadvantages and drawbacks of VES fluids are their high costs, their low tolerance to salts, and stability against high temperatures as found in deep-well applications. However there are recent formulations that overcome these difficulties, at least to some extent.

The components of a viscoelastic fluid are a zwitterionic surfactant, erucyl amidopropyl betaine, an anionic polymer, or *N*-erucyl-*N*,*N*-bis (2-hydroxyethyl)-*N*-methyl ammonium chloride, poly(naphthalene sulfonate), and cationic surfactants, methyl poly(oxyethylene) octadecanammonium chloride and poly(oxyethylene) cocoalkylamines (Couillet and Hughes, 2008; Li et al., 2010). The corresponding fluids exhibit a good viscosity performance.

Typical viscoelastic surfactants are *N*-erucyl-*N*,*N*-bis(2-hydroxyethyl)-*N*-methyl ammonium chloride and potassium oleate, solutions of which form gels when mixed with corresponding activators such as sodium salicylate and potassium chloride (Jones and Tustin, 2010).

The cationic surfactant should be soluble in both organic and inorganic solvents. Solubility in hydrocarbon solvents is promoted by attaching multiple long chain alkyl groups to the active surfactant unit (Jones and Tustin, 2010). Examples are hexadecyltributylphosphonium and trioctylmethylammonium ions. In contrast, cationic surfactants have a single, long, linear hydrocarbon moiety attached to the surfactant group.

In contrast, cationic surfactants for viscoelastic solutions have rather a single, long, linear hydrocarbon moiety attached to the surfactant group. Obviously, there is a conflict between the structural requirements for achieving solubility in hydrocarbons and for the formation of viscoelastic solutions.

As a compromise, surfactant compounds that are suitable for reversibly thickening water-based wellbore fluids and also soluble in both organic and aqueous fluids have been designed.

Tallowamido propylamine oxide (McElfresh and Williams, 2007) is a suitable, nonionic surfactant gelling agent. Nonionic fluids are inherently less damaging to the producing formations than cationic fluid types, and are more efficacious than anionic gelling agents.

The synthesis of branched oleates has been described. 2-Methyl oleic acid methyl ester, or 2-methyl oleate, can be prepared from methyl oleate and methyl iodide in the presence of a pyrimidine-based catalyst (Jones and Tustin, 2010). The methyl ester is then hydrolyzed to obtain 2-methyl oleic acid.

It is sometimes believed that contact with a VES-gelled fluid instantaneously reduces the viscosity of the gel, but it has been discovered that mineral oil can be used as an internal breaker for VES-gelled fluid systems (Crews et al., 2010). The rate of viscosity breaking at a given temperature is influenced by the type and amount of salts present. In the case of low-molecular-weight mineral oils, it is important to add them after the VES component is added to the aqueous fluid.

By using combinations of internal breakers, both the initial and final break of the VES fluid may be customized. Fatty acid compounds or bacteria may be used in addition to mineral oil (Crews, 2006; Crews et al., 2010).

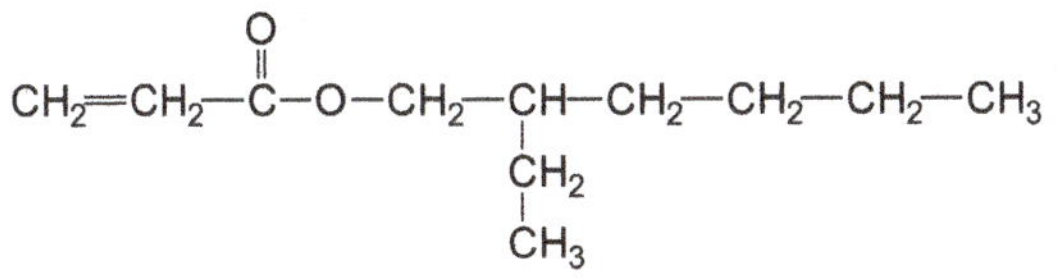

$$CH_2{=}CH_2{-}\overset{O}{\overset{\|}{C}}{-}O{-}CH_2{-}\underset{\underset{CH_3}{|}}{\underset{CH_2}{|}}{CH}{-}CH_2{-}CH_2{-}CH_2{-}CH_3$$

2-Ethylhexylacrylate

FIGURE 3.7 2″-Ethylhexyl acrylate.

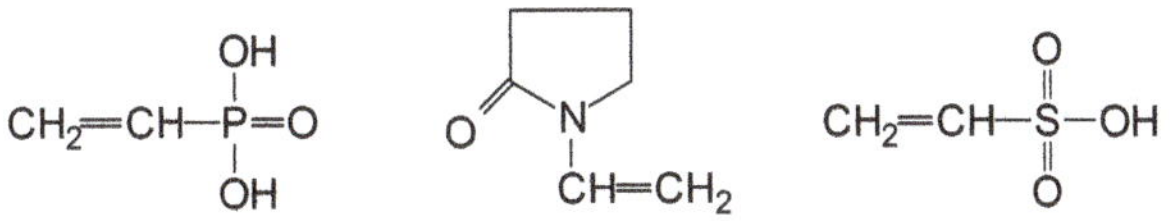

FIGURE 3.8 Monomers for synthetic thickeners.

Miscellaneous Polymers

A copolymer of 2″-ethylhexyl acrylate and acrylic acid is not soluble either in water or in hydrocarbons. The ester units are hydrophobic and the acid units are hydrophilic. An aqueous suspension with a particle size smaller than 10 μm can be useful in preparing aqueous hydraulic fracturing fluids (Harms and Norman, 1988). 2″-Ethylhexyl acrylate is shown in Figure 3.7.

A water-soluble polymer of *N*-vinyl lactam or vinyl-containing sulfonate monomers reduces the water loss and enhances other properties of well-treating fluids in high-temperature subterranean environments (Bharat, 1990). Lignites, tannins, and asphaltic materials are added as dispersants. Vinyl monomers are shown in Figure 3.8.

Lactide Polymers

Degradable thermoplastic lactide polymers are used for fracturing fluids. Hydrolysis is the primary mechanism for degradation of the lactide polymer (Cooke, 2009).

Biodegradable Formulations

Biodegradable drilling fluid formulations have been suggested, which consist of a polysaccharide in a concentration that is insufficient to permit contamination by bacteria. The polymer is a high-viscosity CMC that is sensitive to bacterial enzymes produced by the degradation of the polysaccharide (Pelissier and Biasini, 1991).

The biodegradability of seven kinds of mud additives was studied by determining the content of dissolved oxygen in water, a simple biochemical oxygen demand testing method. The biodegradability is high for starch but lower for polymers of allyl monomers and additives containing an aromatic group (Guo et al., 1996).

CONCENTRATES

Historically, fracture stimulation treatments have been performed by using conventional batch mix techniques, which involves premixing chemicals into tanks and circulating the fluids until a desired gelled fluid rheology is obtained. This method is time-consuming and burdens the oil company with disposal of the fluid if the treatment ends prematurely.

Environmental damage during spillage or disposal can be avoided if the fluid is capable of being gelled as needed. Thus gelling-as-needed technology has been developed with water, methanol, and oil (Gregory et al., 1991). This procedure eliminates batch mixing and minimizes handling of chemicals and base fluid. The customer is charged only for products used, and environmental concerns regarding disposal are virtually eliminated. Computerized chemical addition and monitoring, combined with on-site procedures, ensure quality control throughout treatment. Fluid rheologies can be accurately varied during the treatment by varying polymer loading.

The use of a diesel-based concentrate with hydroxypropyl guar gum has been evolved from batch-mixed dry powder procedures (Harms et al., 1988). The application of such a concentrate reduces system requirements, and companies can benefit from the reduced logistic burden that comes from using the diesel hydroxypropyl guar gum concentrate.

A fracturing fluid slurry concentrate has been proposed (Brannon, 1988) that consists of the components shown in Table 3.4. Such a polymer slurry concentrate will readily disperse and hydrate when admixed with water at the proper pH, thus producing a high-viscosity aqueous fracturing fluid. The fracturing fluid slurry concentrate is useful for producing large volumes of high-viscosity treating fluids at the well site on a continuous basis. Suitable surfactants are shown in Figure 3.9.

Fluidized aqueous suspensions of 15% or more of HEC, hydrophobically modified cellulose ether, hydrophobically modified HEC, methyl cellulose,

TABLE 3.4 Components of a Slurry Concentrate (Brannon, 1988)

Component	Example
Hydrophobic solvent base	Diesel
Suspension agent	Organophilic clay
Surfactant	Ethoxylated nonyl phenol
Hydratable polymer	Hydroxypropyl guar gum

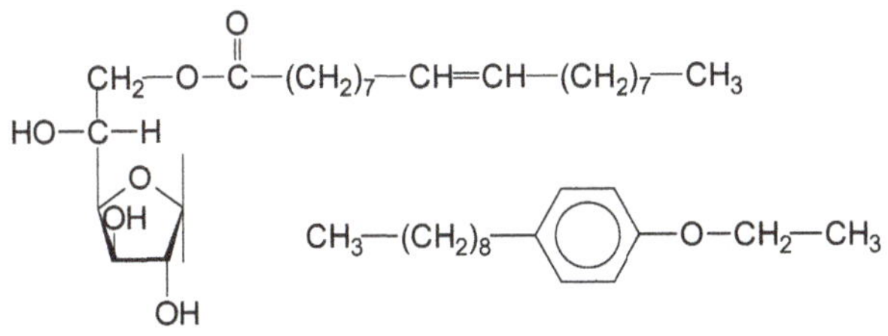

Sorbitan monooleate Ethoxylated nonylphenol

FIGURE 3.9 Surfactants in a slurry concentrate.

hydroxypropylmethyl cellulose, and poly(ethylene oxide) are prepared by adding the polymer to a concentrated sodium formate solution containing xanthan gum as a stabilizer (Burdick and Pullig, 1993).

The xanthan gum is dissolved in water before sodium formate is added. Then the polymer is added to the solution to form a fluid suspension of the polymers. The polymer suspension can serve as an aqueous concentrate for further use.

THICKENERS FOR OIL-BASED SYSTEMS

Organic Gel Aluminum Phosphate Ester

A gel of diesel or crude oil can be produced using a phosphate diester or an aluminum compound with phosphate diester (Gross, 1987). The metal phosphate diester may be prepared by reacting a triester with phosphorous pentoxide to produce a poly(phosphate), which is then reacted with hexanol to produce a phosphate diester (Huddleston, 1989).

The latter diester is then added to the organic liquid along with a nonaqueous source of aluminum, such as aluminum isopropoxide, c.f., Figure 3.10, in diesel oil, to produce the metal phosphate diester. The conditions in the previous reaction steps are controlled to provide a gel with good viscosity versus temperature and time characteristics. All the reagents are substantially free of water and will not affect the pH. The synthesis of a phosphate diesters is

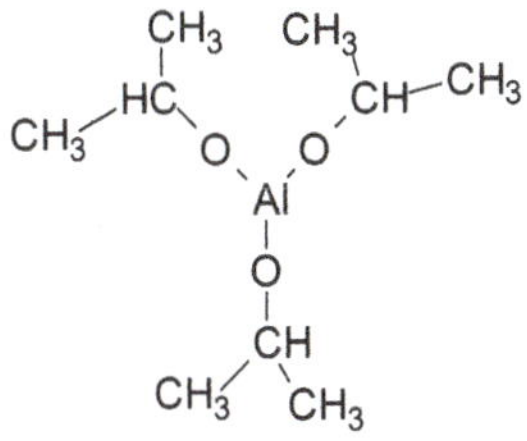

FIGURE 3.10 Aluminum isopropoxide.

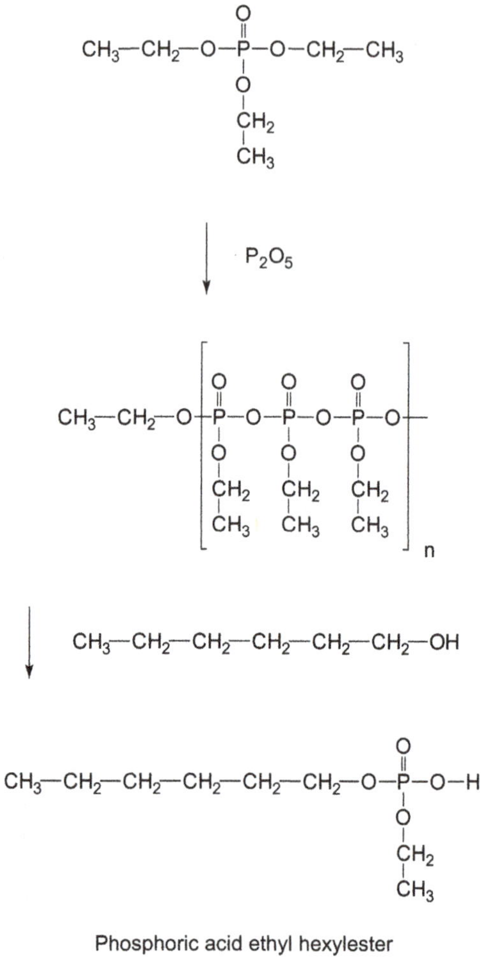

$CH_3-CH_2-O-P(=O)(-O-CH_2-CH_3)-O-CH_2-CH_3$

P_2O_5

$CH_3-CH_2-O-[P(=O)(O-CH_2-CH_3)-O-P(=O)(O-CH_2-CH_3)-O-P(=O)(O-CH_2-CH_3)-O]_n-$

$CH_3-CH_2-CH_2-CH_2-CH_2-CH_2-OH$

$CH_3-CH_2-CH_2-CH_2-CH_2-CH_2-O-P(=O)(O-CH_2-CH_3)-O-H$

Phosphoric acid ethyl hexylester

FIGURE 3.11 Synthesis of higher phosphoric esters.

shown in Figure 3.11. It runs via triethyl phosphate, phosphorous pentoxide, and the esterification reaction with hexanol.

Enhancers for phosphate esters are amino compounds (Geib, 2002). The 2-ethylhexanoic acid trialuminum salt has been suggested with fatty acids as an activator (Subramanian et al., 2001b).

Another method to produce oil-based hydrocarbon gels is to use ferric salts (Smith and Persinski, 1995) rather than aluminum compounds for combination with orthophosphate esters. The ferric salt has the advantage of being usable in the presence of large amounts of water, up to 20%. Ferric salts can be applied in wide ranges of pH. The linkages that are formed can still be broken with gel breaking additives conventionally used for that purpose.

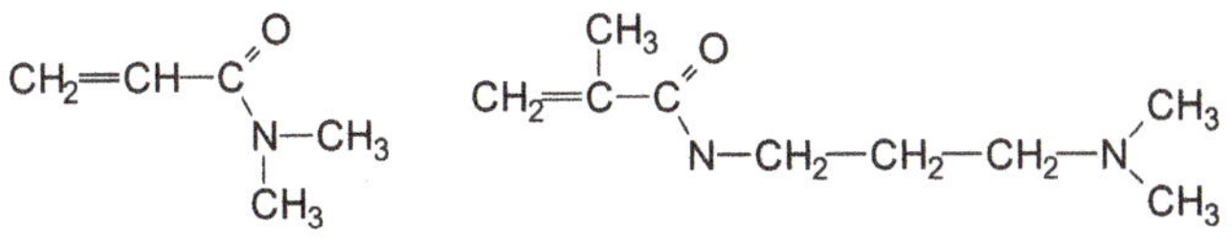

N,N-Dimethylacrylamide Dimethylaminopropyl methacrylamide

FIGURE 3.12 Monomers in a copolymer for viscosifying diesel.

Increasing the Viscosity of Diesel

A copolymer of *N,N*-dimethylacrylamide and *N,N*-dimethyl aminopropyl methacrylamide, a monocarboxylic acid, and ethanolamine may serve to increase the viscosity of diesel or kerosene (Holtmyer and Hunt, 1988). These compounds are shown in Figure 3.12.

VISCOELASTICITY

Viscoelastic materials that exhibit both viscous and elastic properties under mechanical deformation. Such materials exhibit a hysteresis in the stress strain curves. Further a relaxation of stress occurs under constant strain, i.e., the stress is diminished. Moreover, viscoelastic materials are creeping.

A simple model for such materials has been worked out by James Clerk Maxwell already in 1867 (Maxwell, 1867). A Maxwell fluid, or Maxwell body, can be modeled by an idealized viscous damper and by an idealized elastic spring that is connected in series. The basic device is shown in Figure 3.13.

Thus the Maxwell model can explain both elastic properties and viscous properties. Some basic issues of the Maxwell model have been revisited (Rao and Rajagopal, 2007). The basic Maxwell model can be represented by

$$\frac{d\varepsilon_t}{dt} = \frac{d\varepsilon_d}{dt} + \frac{d\varepsilon_s}{dt} = \frac{\sigma}{\eta} + \frac{1}{E}\frac{d\sigma}{dt}. \tag{3.1}$$

The change of total elongation ε_t in time t is the sum of the change of elongation of the damper ε_d and the change of elongation of the spring ε_t. The right-hand side of Eq. (3.1) contains simply Newton's law of flow and Hooke's law of the

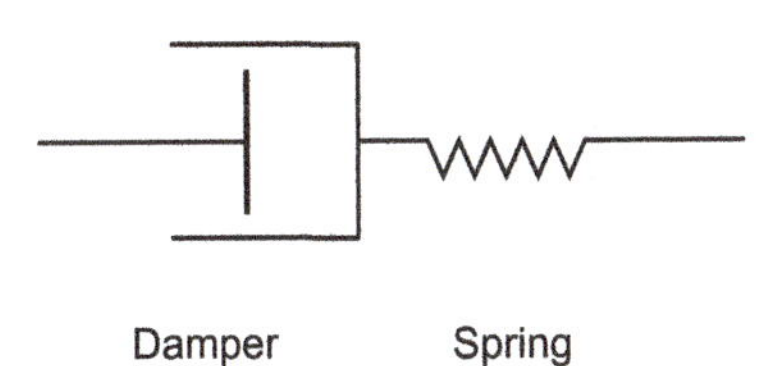

FIGURE 3.13 Maxwell model.

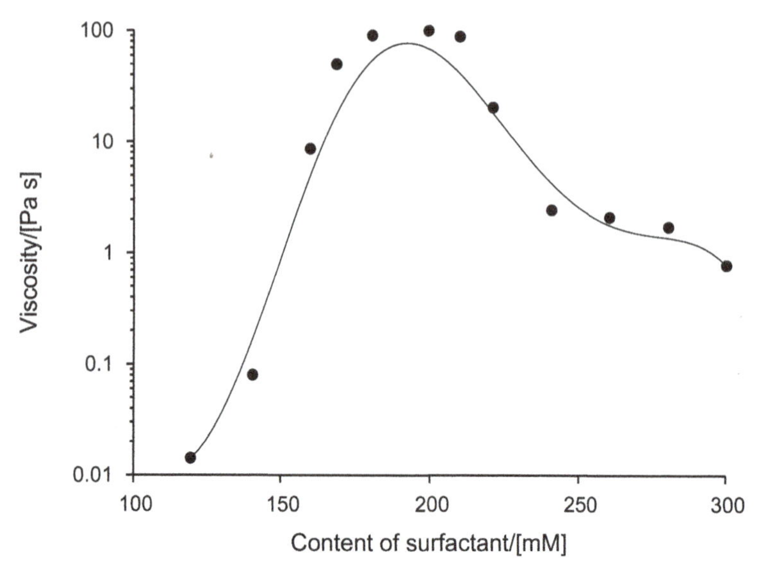

FIGURE 3.14 Viscosity versus concentration of surfactant (Yin et al., 2009).

elongation of an ideal spring, i.e.,

$$\frac{d\varepsilon_d}{dt} = \frac{\sigma}{\eta}, \qquad \varepsilon_s = \frac{1}{E}\sigma,$$

the latter in the rather uncommon differential form. Thus, η is the Newtonian viscosity, σ is the stress, and E is the elastic modulus. By the way, if the spring and the damper are coupled parallel instead of consecutively, the Kelvin-Voigt model emerges, which may be experienced in self-closing doors.

The property of viscoelasticity has been well investigated (Rehage and Hoffmann, 1988). Aqueous solutions of cationic surfactants with strong binding forces of the ions show properties similar to gels.

Microstructural transitions and rheological properties of viscoelastic solutions formed in a catanionic surfactant system were studied using a combination of rheology and dynamic light scattering (Yin et al., 2009). The rheological behavior can be very complicated. In a study on a surfactant system based on dodecyltriethylammonium bromide and sodium dodecyl sulfate, worm-like micelles began to form above a certain surfactant concentration, caused by the growth of small cylindrical micelles. In an intermediate concentration range, the system exhibits a linear viscoelasticity with the characteristics of a Maxwell fluid. Eventually, at higher surfactant concentrations, a transition from linear micelles to branched structures may take place.

The changes of the viscosity at zero shear rate with the total concentration of surfactant with a ratio of dodecyltriethylammonium bromide to sodium dodecyl sulfate of 27/73 are shown in Figure 3.14.

Viscoelastic materials that exhibit both viscous and elastic properties under mechanical deformation. Such materials exhibit a hysteresis in the stress strain curves. Further a relaxation of stress occurs under constant strain, i.e., the stress is diminished. Moreover, viscoelastic materials are creeping.

Viscoelastic Thickeners

Viscoelasticity is caused by a different type of micelle formation than the usual spherical micelles formed by most surfactants. VESs are believed to impart viscosity to an aqueous fluid by the molecules organizing themselves into micelles. Spherical micelles do not give increased viscosity, however, when the micelles have an elongated configuration, for instance are rod-shaped or worm-shaped, they become entangled with one another thereby increasing the viscosity of the fluid (Crews and Huang, 2010).

Elongated VES structures are referred to as living because there is a continuous exchange of VES-type surfactants leaving the VES micelle structures to the aqueous solution and other surfactants leaving the aqueous solution and entering or reentering the VES micelle structures.

VES solutions exhibit a shear thinning behavior. They remain stable despite repeated high-shear applications. By comparison, a typical polymeric thickener will irreversibly degrade when subjected to high-shear applications (Colaco et al., 2007).

Internal breakers work by the rearrangement of the VES micelle from rod-shaped or worm-shaped elongated structures to spherical structures. In other words, they perform the collapse or rearrangement of the viscous elongated micelle structures to nonviscous, more spherical micelle structures (Crews and Huang, 2010).

Enhanced Shear Recovery Agents

Some viscoelastic surfactant fluids exhibit a low shear recovery when subjected to high shear. However, unacceptably long shear recovery times hinder the operation of deep-well oilfield applications. Enhanced shear recovery agents reduce the shear recovery time of a viscoelastic surfactant fluid.

Enhanced shear recovery agents are based on alkylated poly(glucoside)s, poly(glucamide)s, or on copolymers based on ethylene glycol ethyl ether acrylate. Glucamides consist of cyclic forms of glucose in which the hydrogen of the hemiacetal group has been replaced with an alkyl or aryl moiety. The basic structures of glucoside and glucamide are shown in Figure 3.15.

Viscoelastic surfactant fluids are made by mixing suitable surfactants. When the surfactant concentration significantly exceeds a critical level, and eventually subjected to the presence of an electrolyte, the surfactant molecules aggregate and form structures, such as micelles that can interact to form a network exhibiting viscoelastic behavior (Sullivan et al., 2006).

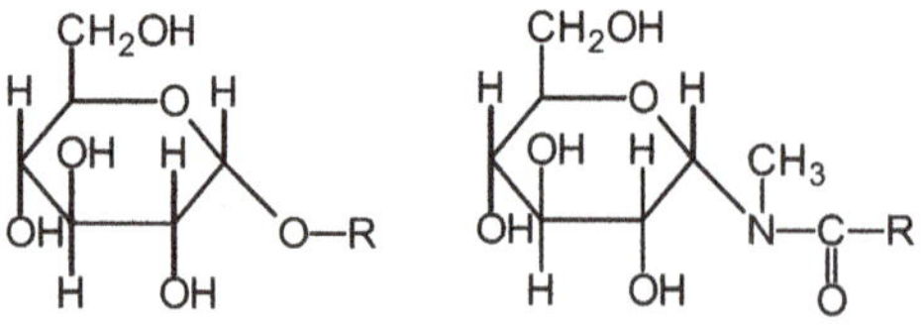

Glucoside Glucamide

FIGURE 3.15 Glucoside and glucamide structures (Colaco et al., 2007).

Viscoelastic surfactant solutions can be formed by the addition of certain reagents to concentrated solutions of surfactants. The surfactants are long chain quaternary ammonium salts such as cetyltrimethyl ammonium bromide.

Tradenames in References

Tradename	Supplier
Description	
Alkaquat™ DMB-451	Rhodia Canada Inc.
Dimethyl benzyl alkyl ammonium chloride (Lawrence and Warrender, 2010)	
Benol®	Sonneborn Refined Products
White mineral oil (Crews and Huang, 2010)	
Captivates® liquid	ISP Hallcrest
Fish gelatin and gum acacia encapsulation coating (Crews and Huang, 2010)	
Carnation®	Sonneborn Refined Products
White mineral oil (Crews and Huang, 2010)	
ClearFRAC™	Schlumberger Technology Corp.
Stimulating Fluid (Couillet and Hughes, 2008; Crews, 2006; Crews and Huang, 2010)	
Diamond FRAQ™	Baker Oil Tools
VES breaker (Crews and Huang, 2010)	
DiamondFRAQ™	Baker Oil Tools
VES System (Crews and Huang, 2010)	
Escaid® (Series)	Crompton Corp.
Mineral oils (Crews and Huang, 2010)	
Geltone® (Series)	Halliburton Energy Services, Inc.
Organophilic clay (Lawrence and Warrender, 2010)	
Hydrobrite® 200	Sonneborn Inc.
White mineral oil (Crews and Huang, 2010)	
Isopar® (Series)	Exxon
Isoparaffinic solvent (Crews and Huang, 2010)	

continued

Tradenames in References (*continued*)

Tradename	Supplier
Description	
Poly-S.RTM	Scotts Comp.
Polymer encapsulation coating (Crews, 2006; Crews and Huang, 2010)	
Rhodafac® LO-11A	Rhodia Inc. Corp.
Phosphate ester (Lawrence and Warrender, 2010)	
Span® 20	Uniqema
Sorbitan monolaurate (Crews and Huang, 2010)	
Span® 40	Uniqema
Sorbitan monopalmitate (Crews and Huang, 2010)	
Span® 65	Uniqema
Sorbitan tristearate (Crews and Huang, 2010)	
Span® 80	Uniqema
Sorbitan monooleate (Crews and Huang, 2010)	
Span® 85	Uniqema
Sorbitan trioleate (Crews and Huang, 2010)	
Tween® 20	Uniqema
Sorbitan monolaurate (Crews and Huang, 2010)	
Tween® 21	Uniqema
Sorbitan monolaurate (Crews and Huang, 2010)	
Tween® 40	Uniqema
Sorbitan monopalmitate (Crews and Huang, 2010)	
Tween® 60	Uniqema
Sorbitan monostearate (Crews and Huang, 2010)	
Tween® 61	Uniqema
Sorbitan monostearate (Crews and Huang, 2010)	
Tween® 65	Uniqema
Sorbitan tristearate (Crews and Huang, 2010)	
Tween® 81	Uniqema
Sorbitan monooleate (Crews and Huang, 2010)	
Tween® 85	Uniqema
Sorbitan monooleate (Crews and Huang, 2010)	
VES-STA 1	Baker Oil Tools
Gel stabilizer (Crews and Huang, 2010)	
WS-44	Halliburton Energy Services, Inc.
Emulsifier (Welton et al., 2010)	

REFERENCES

Bahamdan, A., Daly, W.H., 2007. Hydrophobic guar gum derivatives prepared by controlled grafting processes. Polym. Adv. Technol. 18 (8), 652–659. http://dx.doi.org/10.1002/pat.874.

Baranet, S.E., Hodge, R.M., Kucera, C.H., 1987. Stabilized fracture fluid and crosslinker therefor. US Patent 4686052, assigned to Dowell Schlumberger Inc., Tulsa, OK, 11 August 1987. <http://www.freepatentsonline.com/4686052.html>.

Batelaan, J.G., van der Horts, P.M., 1994. Method of making amide modified carboxyl-containing polysaccharide and fatty amide-modified polysaccharide so obtainable. WO Patent 9424169, assigned to Akzo Nobel NV, 27 October 1994.

Bharat, P., 1990. Well treating fluids and additives therefor. EP Patent 372469, 13 June 1990.

Brannon, H.D., 1988. Fracturing fluid slurry concentrate and method of use. EP Patent 280341, 31 August 1988.

Burba, J.L.I., Strother, G.W., 1991. Mixed metal hydroxides for thickening water or hydrophylic fluids. US Patent 4990268, assigned to Dow Chemical Co., 5 February 1991.

Burba, J.L.I., Hoy, E.F., Read Jr., A.E., 1992. Adducts of clay and activated mixed metal oxides. WO Patent 9218238, assigned to Dow Chemical Co., 29 October 1992.

Burdick, C.L., Pullig, J.N., 1993. Sodium formate fluidized polymer suspensions process. US Patent 5228908, assigned to Aqualon Co., 20 July 1993.

Colaco, A., Marchand, J.-P., Li, F., Dahanayake, M.S., 2007. Viscoelastic surfactant fluids having enhanced shear recovery, rheology and stability performance. US Patent 7279446, assigned to Rhodia Inc., Cranbury, NJ, Schlumberger Technology Corp., Sugarland, TX, 9 October 2007. <http://www.freepatentsonline.com/7279446.html>.

Cooke Jr., C.E., 2009. Method and materials for hydraulic fracturing of wells using a liquid degradable thermoplastic polymer. US Patent 7569523, 4 August 2009. <http://www.freepatentsonline.com/7569523.html>.

Couillet, I., Hughes, T., 2008. Aqueous fracturing fluid. US Patent 7427583, assigned to Schlumberger Technology Corporation, Ridgefield, CT, 23 September 2008. <http://www.freepatentsonline.com/7427583.html>.

Courtois-Sambourg, J., Courtois, B., Heyraud, A., Colin-Morel, P., Rinaudo-Duhem, M., 1993. Polymer compounds of the glycuronic acid, method of preparation and utilization particularly as gelifying, thickening, hydrating, stabilizing, chelating or flocculating means. WO Patent 9318174, assigned to Picardie Univ., 16 September 1993.

Crews, J.B., 2006. Bacteria-based and enzyme-based mechanisms and products for viscosity reduction breaking of viscoelastic fluids. US Patent 7052901, assigned to Baker Hughes Inc., Houston, TX, 30 May 2006. <http://www.freepatentsonline.com/7052901.html>.

Crews, J.B., Huang, T., 2010. Use of oil-soluble surfactants as breaker enhancers for ves-gelled fluids. US Patent 7696135, assigned to Baker Hughes Inc., Houston, TX, 13 April 2010. <http://www.freepatentsonline.com/7696135.html>.

Crews, J.B., Huang, T., Gabrysch, A.D., Treadway, J.H., Willingham, J.R., Kelly, P.A., Wood, W.R., 2010. Methods and compositions for fracturing subterranean formations. US Patent 7723272, assigned to Baker Hughes Inc., Houston, TX, 25 May 2010. <http://www.freepatentsonline.com/7723272.html>.

Deem, C.K., Schmidt, D.D., Molner, R.A., 1991. Use of MMH (mixed metal hydroxide)/propylene glycol mud for minimization of formation damage in a horizontal well. In: Proceedings Volume. No. 91–29. 4th Cade/Caodc Spring Drilling Conf., Calgary, Canada, 10–12 April 1991, Proc.

Doherty, D.H., Ferber, D.M., Marrelli, J.D., Vanderslice, R.W., Hassler, R.A., 1992. Genetic control of acetylation and pyruvylation of xanthan based polysaccharide polymers. WO Patent 9219753, assigned to Getty Scientific Dev Co., 12 November 1992.

Doolan, J.G., Cody, C.A., 1995. Pourable water dispersible thickening composition for aqueous systems and a method of thickening said aqueous systems. US Patent 5425806, assigned to Rheox Inc., 20 June 1995.

Egraz, J.B., Grondin, H., Suau, J.M., 1994. Acrylic copolymer partially or fully soluble in water, cured or not and its use (copolymere acrylique partiellement ou totalement hydrosoluble, reticule ou non et son utilisation). EP Patent 577526, assigned to Coatex SA, 5 January 1994.

Felixberger, J., 1996. Mixed metal hydroxides (MMH)—an inorganic thickener for water-based drilling muds (Mixed Metal Hydroxide (MMH)—Ein anorganisches Verdickungsmittel für wasserbasierte Bohrspülungen). In: Proceedings Volume. DMGK Spring Conf., Celle, Germany, 25–26 April 1996, pp. 339–351.

Geib, G.G., 2002. Hydrocarbon gelling compositions useful in fracturing formations. US Patent 6342468, assigned to Ethox Chemicals Llc., 29 January 2002.
Gotlieb, K.F., Bleeker, I.P., van Doren, H.A., Heeres, A., 1996. 2-nitroalkyl ethers of native or modified starch, method for the preparation thereof, and ethers derived therefrom. EP Patent 710671, assigned to Coop Verkoop Prod. Aard De, 8 May 1996.
Gregory, G., Shuell, D., Thompson Sr., J.E., 1991. Overview of contemporary lfc (liquid frac concentrate) fracture treatment systems and techniques. In: Proceedings Volume. No. 91–01. 4th Cade/caodc Spring Drilling Conf., Calgary, Canada, 10–12 April 1991, Proc.
Gross, J.M., 1987. Gelling organic liquids. EP Patent 225661, assigned to Dowell Schlumberger Inc., 16 June 1987.
Guo, D.R., Gao, J.P., Lu, K.H., Sun, M.B., Wang, W., 1996. Study on the biodegradability of mud additives. Drilling Fluid and Completion Fluid 13 (1), 10–12.
Harms, W.M., Norman, L.R., 1988. Concentrated hydrophilic polymer suspensions. US Patent 4772646, 20 September 1988.
Harms, W.M., Watts, M., Venditto, J., Chisholm, P., 1988. Diesel-based hpg (hydroxypropyl guar) concentrate is product of evolution. Pet. Eng. Int. 60 (4), 51–54.
Hodge, R.M., 1997. Particle transport fluids thickened with acetylate free xanthan heteropolysaccharide biopolymer plus guar gum. US Patent 5591699, assigned to Du Pont De Nemours andand Co., 7 January 1997.
Holtmyer, M.D., Hunt, C.V., 1988. Method and composition for viscosifying hydrocarbons. US Patent 4780221, 25 October 1988.
Holtmyer, M.D., Hunt, C.V., 1992. Crosslinkable cellulose derivatives. EP Patent 479606, assigned to Halliburton Co., 8 April 1992.
House, R.F., Cowan, J.C., 2001. Chitosan-containing well drilling and servicing fluids. US Patent 6258755, assigned to Venture Innovations Inc., 10 July 2001.
Huddleston, D.A., 1989. Hydrocarbon geller and method for making the same. US Patent 4877894, assigned to Nalco Chemical Co., 31 October 1989.
Jafry, H.R., Pasquali, M., Barron, A.R., 2011. Effect of functionalized nanomaterials on the rheology of borate cross-linked guar gum. Ind. Eng. Chem. Res. 50 (6), 3259–3264. http://dx.doi.org/10.1021/ie101836z.
Jones, T.G.J., Tustin, G.J., 2010. Process of hydraulic fracturing using a viscoelastic wellbore fluid. US Patent 7655604, assigned to Schlumberger Technology Co., Ridgefield, CT, 2 February 2010. <http://www.freepatentsonline.com/7655604.html>.
Keilhofer, G., Plank, J., 2000. Solids composition based on clay minerals and use thereof. US Patent 6025303, assigned to Skw Trostberg AG, 15 February 2000.
Kelly, P.A., Gabrysch, A.D., Horner, D.N., 2007. Stabilizing crosslinked polymer guars and modified guar derivatives. US Patent 7195065, assigned to Baker Hughes Inc., Houston, TX, 27 March 2007. <http://www.freepatentsonline.com/7195065.html>.
Kotelnikov, V.S., Demochko, S.N., Melnik, M.P., Mikitchak, V.P., 1992. Improving properties of drilling solution—by addition of ferrochrome- lignosulphonate and aqueous solution of cement and carboxymethyl cellulose. SU Patent 1730118, assigned to Ukr. Natural Gas Res. Inst., 30 April 1992.
Kucera, C.H., 1987. Fracturing of subterranean formations. US Patent 4683068, assigned to Dowell Schlumberger Inc., Tulsa, OK, 28 July 1987. <http://www.freepatentsonline.com/4683068.html>.
Lange, P., Plank, J., 1999. Mixed metal hydroxide (MMH) viscosifier for drilling fluids: Properties and mode of action (Mixed Metal Hydroxide (MMH)—Eigenschaften und Wirkmechanismus als Verdickungsmittel in Bohrspülungen). Erdöl Erdgas Kohle 115 (7–8), 349–353.
Langlois, B., 1998. Fluid comprising cellulose nanofibrils and its use for oil mining. WO Patent 9802499, assigned to Rhone Poulenc Chimie, 22 January 1998.
Langlois, B., Guerin, G., Senechal, A., Cantiani, R., Vincent, I., Benchimol, J., 1999. Fluid comprising cellulose nanofibrils and its use for oil mining (fluide comprenant des nanofibrilles de cellulose et son application pour l'exploitation de gisements petroliers). EP Patent 912653, assigned to Rhodia Chimie, 6 May 1999.
Lawrence, S., Warrender, N., 2010. Crosslinking composition for fracturing fluids. US Patent 7749946, assigned to Sanjel Co., Calgary, Alberta, CA, 6 July 2010. <http://www.freepatentsonline.com/7749946.html>.
Lemanczyk, Z.R., 1992. The use of polymers in well stimulation: An overview of application, performance and economics. Oil Gas Europe Mag. 18 (3), 20–26.

Li, F., Dahanayake, M., Colaco, A., 2010. Multicomponent viscoelastic surfactant fluid and method of using as a fracturing fluid. US Patent 7 772 164, assigned to Rhodia, Inc., Cranbury, NJ, 10 August 2010. <http://www.freepatentsonline.com/7772164.html>.
Lukach, C.A., Zapico, J., 1994. Thermally stable hydroxyethylcellulose suspension. EP Patent 619 340, assigned to Aqualon Co., 12 October 1994.
Lundan, A.O., Lahteenmaki, M.J., 1996. Stable cmc (carboxymethyl cellulose) slurry. US Patent 5 487 777, assigned to Metsa Serla Chemicals Oy, 30 January 1996.
Lundan, A.O., Anas, P.H., Lahteenmaki, M.J., 1993. Stable cmc (carboxymethyl cellulose) slurry. WO Patent 9 320 139, assigned to Metsa Serla Chemicals Oy, 14 October 1993.
Maxwell, J.C., 1867. On the dynamical theory of gases. Phil. Trans. R. Soc. Lond. 157 (1), 49–88.
McElfresh, P.M., Williams, C.F., 2007. Hydraulic fracturing using non-ionic surfactant gelling agent. US Patent 7 216 709, assigned to Akzo Nobel N.V., Arnhem, NL, 15 May 2007. <http://www.freepatentsonline.com/7216709.html>.
Meyer, V., Audibert-Hayet, A., Gateau, J.P., Durand, J.P., Argillier, J.F., 1999. Water-soluble copolymers containing silicon. GB Patent 2 327 946, assigned to Inst. Francais Du Petrole, 10 February 1999.
Mondshine, T.C., 1987. Crosslinked fracturing fluids. WO Patent 8 700 236, assigned to Texas United Chemical Corp., 15 January 1987.
Mueller, H., Breuer, W., Herold, C.P., Kuhm, P., von Tapavicza, S., 1997. Mineral additives for setting and/or controlling the rheological properties and gel structure of aqueous liquid phases and the use of such additives. US Patent 5 663 122, assigned to Henkel KG Auf Aktien, 2 September 1997.
Patel, B.B., 1994. Fluid composition comprising a metal aluminate or a viscosity promoter and a magnesium compound and process using the composition. EP Patent 617 106, assigned to Phillips Petroleum Co., 28 September 1994.
Pelissier, J.J.M., Biasini, S., 1991. Biodegradable drilling mud (boue de forage biodegradable). FR Patent 2 649 988, 25 January 1991.
Putzig, D.E., 2012. Zirconium-based cross-linking composition for use with high pH polymer solutions. US Patent 8 153 564, assigned to Dorf Ketal Speciality Catalysts, LLC, Stafford, TX, 10 April 2012. <http://www.freepatentsonline.com/8153564.html>.
Rangus, S., Shaw, D.B., Jenness, P., 1993. Cellulose ether thickening compositions. WO Patent 9 308 230, assigned to Laporte Industries Ltd., 29 April 1993.
Rao, I.J., Rajagopal, K.R., 2007. On a new interpretation of the classical Maxwell model. Mech. Res. Commun. 34 (7–8), 509–514. <http://www.sciencedirect.com/science/article/B6V48-4P6M5XY-3/2/55d0b1078e2f1113c002c004729d5079>.
Rehage, H., Hoffmann, H., 1988. Rheological properties of viscoelastic surfactant systems. J. Phys. Chem. 92 (16), 4712–4719.
Robinson, F., 1996. Polymers useful as pH responsive thickeners and monomers therefor. WO Patent 9 610 602, assigned to Rhone Poulenc Inc., 11 April 1996.
Robinson, F., 1999. Polymers useful as pH responsive thickeners and monomers therefor. US Patent 5 874 495, assigned to Rhodia, 23 February 1999.
Rummo, G.J., Startup, R., 1986. Bisalkyl bis(trialkanol amine)zirconates and use of same as thickening agents for aqueous polysaccharide solutions. US Patent 4 578 488, assigned to Kay-Fries, Inc., Stony Point, NY, 25 March 1986. <http://www.freepatentsonline.com/4578488.html>.
Smith, K.W., Persinski, L.J., 1995. Hydrocarbon gels useful in formation fracturing. US Patent 5 417 287, assigned to Clearwater Inc., 23 May 1995.
Subramanian, S., Islam, M., Burgazli, C.R., 2001a. Quaternary ammonium salts as thickening agents for aqueous systems. WO Patent 0 118 147, assigned to Crompton Corp., 15 March 2001.
Subramanian, S., Zhu, Y.P., Bunting, C.R., Stewart, R.E., 2001b. Gelling system for hydrocarbon fluids. WO Patent 0 109 482, assigned to Crompton Corp., 8 February 2001.
Sullivan, P., Christanti, Y., Couillet, I., Davies, S., Hughes, T., Wilson, A., 2006. Methods for controlling the fluid loss properties of viscoelastic surfactant based fluids. US Patent 7 081 439, assigned to Schlumberger Technology Corporation, Sugar Land, TX, 25 July 2006. <http://www.freepatentsonline.com/7081439.html>.

Waehner, K., 1990. Experience with high temperature resistant water based drilling fluids (Erfahrungen beim Einsatz hochtemperatur-stabiler wasserbasischer Bohrspülung). Erdöl Erdgas Kohle 106 (5), 200–201.

Welton, T.D., Todd, B.L., McMechan, D., 2010. Methods for effecting controlled break in pH dependent foamed fracturing fluid. US Patent 7662756, assigned to Halliburton Energy Services, Inc., Duncan, OK, 16 February 2010. <http://www.freepatentsonline.com/7662756.html>.

Westland, J.A., Lenk, D.A., Penny, G.S., 1993. Rheological characteristics of reticulated bacterial cellulose as a performance additive to fracturing and drilling fluids. In: Proceedings Volume. SPE Oilfield Chemical International Symposium, New Orleans, 3–5 March 1993, pp. 501–514.

Wilkerson, J.M.I., Verstrat, D.W., Barron, M.C., 1995. Associative monomers. US Patent 5412142, assigned to National Starch and Chemical Investment holding Corp., 2 May 1995.

Yeh, M.H., 1995a. Anionic sulfonated thickening compositions. EP Patent 632057, assigned to Rhone Poulenc Spec. Chem. C, 4 January 1995.

Yeh, M.H., 1995b. Compositions based on cationic polymers and anionic xanthan gum (compositions a base de polymeres cationiques et de gomme xanthane anionique). EP Patent 654482, assigned to Rhone Poulenc Spec. Chem. C, 24 May 1995.

Yin, H., Lin, Y., Huang, J., 2009. Microstructures and rheological dynamics of viscoelastic solutions in a catanionic surfactant system. J. Colloid Interface Sci. 338 (1) 177–183. <http://www.sciencedirect.com/science/article/B6WHR-4WGK6P7-2/2/38daea0593cc7f3234 ba25febeefe035>.

Zeilinger, S.C., Mayerhofer, M.J., Economides, M.J., 1991. A comparison of the fluid-loss properties of borate-zirconate-crosslinked and noncrosslinked fracturing fluids. In: Proceedings Volume. SPE East Reg Conference, Lexington, KY, 23–25 October 1991, pp. 201–209.

Chapter 4

Friction Reducers

During the drilling, completion, and stimulation of subterranean wells, treatment fluids are often pumped through tubular structures, e.g., pipes, or coiled tubing. A considerable amount of energy may be lost due to turbulence in the treatment fluid. As a result of these energy losses, additional horsepower may be necessary to achieve the desired treatment (Robb et al., 2010).

Low pumping friction pressures are achieved by delaying the crosslinking of the compositions, but there are also specific additives available to reduce the drag in the tubings. The first application of drag reducers was using guar in oil well fracturing, now a routine practice.

Relatively small quantities of a bacterial cellulose ($0.60–1.8\,gl^{-1}$) in hydraulic fracturing fluids enhance their rheologic properties (Penny et al., 1991). The suspension of the proppant is enhanced and friction loss through well casings is reduced.

INCOMPATIBILITY

Although quaternary surfactants can be useful when used together with biocides, some quaternary surfactants may have a fundamental incompatibility with anionic friction reducers, which are also useful in subterranean operations.

It is believed that this fundamental incompatibility may arise from charges present on both molecules that may cause the quaternary surfactant and the friction reducer to react and eventually a precipitate is formed. Additionally, some biocides, e.g., such as oxidizers, may degrade certain friction reducers (Bryant et al., 2009).

POLYMERS

Poly(acrylamide) (PAM) and poly(acrylate) polymers and copolymers are used typically for high-temperature applications or friction reducers at low concentrations for all temperature ranges (Lukocs et al., 2007). To reduce the energy losses in the course of pumping, certain friction reducing polymers have been included in treatment fluids.

Hydraulic Fracturing Chemicals and Fluids Technology. http://dx.doi.org/10.1016/B978-0-12-411491-3.00004-2
© 2013 Elsevier Inc. All rights reserved.

TABLE 4.1 Monomers for Friction Reducing Polymers (Robb et al., 2010)

Monomer
Acrylamide
Acrylic acid
2-Acrylamido-2-methylpropane sulfonic acid
N,N-dimethylacrylamide
Vinyl sulfonic acid
N-Vinyl acetamide
N-Vinyl formamide
Itaconic acid
Methacrylic acid
Acrylic acid esters
Methacrylic acid esters

In general, friction reducing polymers are high-molecular-weight polymers, such as those having a molecular weight of at least about 2.5 MDa. Typically, friction reducing polymers are linear and flexible. One example of such a friction reducing polymer is a copolymer from acrylamide (AAm) and acrylic acid (Robb et al., 2010). A wide variety of monomers may be suitable for use with the present technique. These are summarized in Table 4.1.

Friction reducing polymers may be in the acid form or in the salt form. Multivalent ions contained in the water used to prepare the treatment fluids may undesirably interact with the friction reducing polymers. The multivalent ions may reduce the effectiveness of the friction reducing polymers. The use of complexing agents to control the multivalent ions in the water can improve the performance of the friction reducing polymers.

It has been discovered that adding the complexing agent to the concentrated polymer composition rather than to the water may reduce the amount of the complexing agent needed to improve performance of the friction reducing polymers. By adding an inorganic complexing agent to the oil continuous phase, where it is insoluble, a much improved friction reducer performance can be observed. Examples of complexing agents are shown in Table 4.2.

ENVIRONMENTAL ASPECTS

The use of friction reducing polymers has proved challenging from an environment standpoint (King et al., 2007). For example, many of the friction reducing polymers that have been used previously are provided as oil-external emulsion polymers, wherein upon addition to the aqueous treatment fluid, the emulsion should invert, releasing the friction reducing polymer into the fluid.

TABLE 4.2 Complexing Agents (McMechan et al., 2009; Robb et al., 2010)

Compound
Carbonates
Phosphates
Pyrophosphates
Orthophosphates
Citric acid
Gluconic acid
Glucoheptanoic acid
Ethylenediamine tetraacetic acid
Nitrilotriacetic acid

While the aqueous pad fluids and other aqueous well-treating fluids containing friction pressure reducers utilized have been used successfully, the friction pressure reducers have been suspended in a hydrocarbon-water emulsion and as a result they have been toxic and detrimental to the environment. Thus, there are needs for improved friction pressure reducers which are nontoxic and environmentally acceptable (King et al., 2004).

The hydrocarbon carrier fluid present in the oil-external emulsion may pose environmental problems with the subsequent disposal of the treatment fluid. Among other reasons, disposal of hydrocarbons, e.g., such as the carrier fluid in the oil-external emulsion, may have undesirable environmental characteristics or may be limited by strict environmental regulations in certain areas of the world (King et al., 2007).

Furthermore, the hydrocarbon carrier fluid present in the oil-external emulsion also may undesirably contaminate water in the formation. Water-based friction reducing polymers seem to be more suitable than other polymer types for environmental aspects.

A nontoxic environmentally acceptable friction pressure reducer can be formulated from a copolymer of AAm and dimethylaminoethyl acrylate quaternized with benzyl chloride and a stabilizing and dispersing homopolymer of acryloyloxyethyltrimethyl ammonium chloride (King et al., 2004). The structure of this monomer is shown in Figure 4.1.

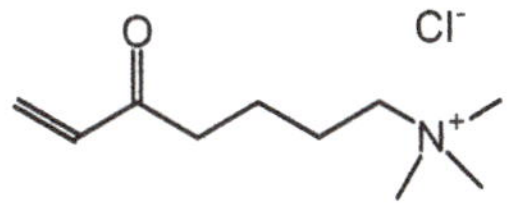

FIGURE 4.1 Acryloyloxyethyltrimethyl ammonium chloride.

TABLE 4.3 Friction Reduction by AMPS-based Polymers for Foamed Liquids (Harris and Heath, 2006)

Polymer type	AMPS/(%)	Friction loss(%)
Acrylamide, traces acrylate	0	0
15–20% AMPS, acrylamide, traces acrylate	15–20	13
Cationic acrylamide	0	0
<10% AMPS, acrylamide, traces acrylate	10	0
Cationic acrylamide	0	0
20–25% AMPS, acrylamide, acrylate	20–25	22
40% AMPS, acrylamide, traces acrylate	40	39

CARBON DIOXIDE FOAMED FLUIDS

The capability of friction reduction of 2-acrylamido-2-methyl-1-propane sulfonic acid (AMPS) on carbon dioxide foamed fluids using 20% carbon dioxide by volume is shown in Table 4.3.

Polymer Emulsions

Copolymers from AAm and an anionic monomer, such as acrylamidopropane sulfonic acid, acrylic acid, methacrylic acid, monoacryloxyethyl phosphate, or the corresponding alkali metal salts, have been described as viscosity reducing components (Parnell et al., 2012).

By adding a relatively large amount of surfactant, the hydrophilicity and dispersibility of the polymer is increased, thus increasing the stability of the system in the fluid. Furthermore, an increased surfactant level increases the inversion rate of the additive, even under low energy conditions. As a result, less polymer is needed to achieve the desired friction reducing performance.

However, inverting surfactants may adversely interact with the emulsifier or emulsion and destroy it prior to use. Therefore polymer emulsions generally contain less than 5% of inverting surfactant. Polymer emulsions with this low amount of inverting surfactant, however, may not provide the desired reduction in friction because the polymer emulsion either does not invert completely or is not brine or acid tolerant.

In the formulations a stepwise addition of two solvent surfactant combinations is used (Parnell et al., 2012). The solvent is preferably a terpene, such as *d*-limonene, dipentene, *l*-limonene, *d,l*-limonene, myrcene, and α-pinene. The surfactant is selected from ethoxylated glycerides, ethoxylated sorbitan esters, ethoxylated alkyl phenols, ethoxylated alcohols, castor oil ethoxylates, cocoamide ethoxylates, and sorbitan monooleates. The surfactant should have a hydrophile-lipophile balance of above about 7.

OIL-EXTERNAL COPOLYMER EMULSIONS

Emulsion polymerization may be used to prepare an oil-external copolymer emulsion to get a friction reducing copolymer. Techniques of emulsion polymerization can be varied with respect to initiation temperature, amount and type of initiator, monomers, and stirring rate. An example of such a composition that may be used for a suitable oil-external copolymer emulsion is shown in Table 4.4.

Friction reduction tests were performed using friction reducing copolymers fabricated from varying concentrations of acrylamide and acrylic acid. The results are shown in Table 4.5.

POLY(ACRYLAMIDE) WITH WEAK LABILE LINKS

The availability of water-soluble polymers containing weak links in the backbone of the polymer that can be degraded upon experiencing a certain trigger, such as temperature, pH, or reducing agent, is highly advantageous (Kot et al., 2012). Because of the ability of weak links to degrade under certain

TABLE 4.4 Composition for Emulsion Polymerization (Chatterji et al., 2006)

Component	Amount (%)
Paraffinic/naphthenic organic solvent	21.1732
Tall oil fatty acid diethanolamine	1.1209
Poly(oxyethylene) sorbitan monooleate	0.0722
Sorbitan monooleate	0.3014
Acrylamide	22.2248
4-Methoxyphenol	0.0303
Ammonium chloride	1.6191
Acrylic acid	4.3343
Ethylenediamine tetracetic acid	0.0237
tert-Butyl hydroperoxide	0.0023
Sodium metabisulfite	0.2936
2,2′-Azobis(2-amidinopropane)dihydrochloride	0.0311
Ethoxylated C_{12}–C_{16} alcohol	1.3700
Water	43.1737

TABLE 4.5 Friction Reduction (Chatterji et al., 2006)

After Time/(min)	Ratio of Acrylamide/Acrylic Acid			
	70/30	85/15	87.5/12.5	90/10
4	65.9	66.3	62.2	57.2
9	61.0	56.1	54.3	50.2
14	55.2	49.8	50.3	45.2
19	50.0	45.8	45.7	41.3
Maximum	69.7	71.1	70.7	69.7

conditions, such polymers can be used for their intended application and can afterwards be degraded in a controlled and predetermined way. The resulting lower-molecular-weight fractions of that polymer lead to a reduced viscosity and quick partitioning into the water phase, and they are also less likely to adsorb onto formation surfaces. Additionally, no oxidizers need to be pumped to break or clean the deposited polymer, thus saving treatment time.

It has been proved that using a bifunctional reducing agent containing degradable groups and oxidizing metal ions as a redox couple is an effective method to initiate free-radical polymerization and build degradable groups into the backbone of vinyl polymers.

Temperature-degradable but hydrolytically stable azo groups showed the most-desirable results. The presence of azo groups in the backbone of the PAM was confirmed by nuclear magnetic-resonance spectroscopy and differential scanning calorimetry. The degradation behavior of the PAM with temperature-sensitive azo groups was characterized using gel permeation chromatography and proved that multiple labile links were built into the polymer backbone. It was also found that PAM with azo links in the polymer backbone is as good a drag-reducing agent as pure PAM. However, PAM with azo links in the backbone loses its drag-reduction properties once subjected to elevated temperatures, which for some applications is viewed as an advantage (Kot et al., 2012).

Tradenames in References

Tradename	Supplier
Description	
Alkamuls® SMO	Rhodia HPCII
Sorbitan monooleate (Chatterji et al., 2006)	
Amadol® 511	Akzo Nobel Surface Chemistry
Tall oil fatty acid diethanolamine (Chatterji et al., 2006)	
FR™ (Series)	Halliburton Energy Services, Inc.
Friction reducer (Robb et al., 2010)	
LPA® –210	Sasol North America, Inc.
Hydrocarbon solvent (Chatterji et al., 2006)	
Surfonic®	Huntsman Performance Products
Emulsifier, ethoxylated C_{12} alcohol (Chatterji et al., 2006)	
Tween® 81	Uniqema
Sorbitan monooleate (Chatterji et al., 2006)	

REFERENCES

Bryant, J.E., McMechan, D.E., McCabe, M.A., Wilson, J.M., King, K.L., 2009. Treatment fluids having biocide and friction reducing properties and associated methods. US Patent Application 20090229827, 17 September 2009. <http://www.freepatentsonline.com/20090229827.html>.

Chatterji, J., King, K.L., McMechan, D.E., 2006. Subterranean treatment fluids, friction reducing copolymers, and associated methods. US Patent 7 004 254, assigned to Halliburton Energy Services, Inc., Duncan, OK, 28 February 2006. <http://www.freepatentsonline.com/7004254.html>.

Harris, P.C., Heath, S.J., 2006. Friction reducers for fluids comprising carbon dioxide and methods of using friction reducers in fluids comprising carbon dioxide. US Patent 7 117 943, assigned to Halliburton Energy Services, Inc., Duncan, OK, 10 October 2006. <http://www.freepatentsonline.com/7117943.html>.

King, K.L., McMechan, D.E., Chatterji, J., 2004. Methods, aqueous well treating fluids and friction reducers therefor. US Patent 6 784 141, assigned to Halliburton Energy Services, Inc., Duncan, OK, 31 August 2004. <http://www.freepatentsonline.com/6784141.html>.

King, K.L., McMechan, D.E., Chatterji, J., 2007. Water-based polymers for use as friction reducers in aqueous treatment fluids. US Patent 7 271 134, assigned to Halliburton Energy Services, Inc., Duncan, OK, 18 September 2007. <http://www.freepatentsonline.com/7271134.html>.

Kot, E., Saini, R., Norman, L.R., Bismarck, A., 2012. Novel drag-reducing agents for fracturing treatments based on polyacrylamide containing weak labile links in the polymer backbone (includes supplementary experimental section). SPE J. 17 (3), 924–930. <http://www.onepetro.org/mslib/app/Preview.do?paperNumber=SPE-141257-PA&societyCode=SPE>.

Lukocs, B., Mesher, S., Wilson, T.P.J., Garza, T., Mueller, W., Zamora, F., Gatlin, L.W., 2007. Non-volatile phosphorus hydrocarbon gelling agent. US Patent Application 20070173413, assigned to Clearwater International, LLC., 26 July 2007. <http://www.freepatentsonline.com/20070173413.html>.

McMechan, D.E., Hanes Jr., R.E., Robb, I.D., Welton, T.D., King, K.L., King, B.J., Chatterji, J., 2009. Friction reducer performance by complexing multivalent ions in water. US Patent 7 579 302, assigned to Halliburton Energy Services, Inc., Duncan, OK, 25 August 2009. <http://www.freepatentsonline.com/7579302.html>.

Parnell, E., Sanner, T., Holtmyer, M., Philpot, D., Zelenev, A., Gilzow, G., Champagne, L., Sifferman, T., 2012. Drag-reducing copolymer compositions. US Patent Application 20120035085, assigned to Cesi Chemical, Inc., Marlow, OK, 9 February 2012. <http://www.freepatentsonline.com/20120035085.html>.

Penny, G.S., Stephens, R.S., Winslow, A.R., 1991. Method of supporting fractures in geologic formations and hydraulic fluid composition for same. US Patent 5009797, assigned to Weyerhaeuser Co., 23 April 1991.

Robb, I.D., Welton, T.D., Bryant, J., Carter, M.L., 2010. Friction reducer performance in water containing multivalent ions. US Patent 7846878, assigned to Halliburton Energy Services, Inc., Duncan, OK, 7 December 2010. <http://www.freepatentsonline.com/7846878.html>.

Chapter 5

Fluid Loss Additives

The basic issues of fluid loss additives have been reviewed (Harris, 1988; Rimassa et al., 2009). Fluid loss additives are also called *filtrate-reducing agents*. Fluid losses may occur when the fluid comes in contact with a porous formation. This is relevant for drilling and completion fluids, fracturing fluids, and cement slurries.

The extent of fluid loss is dependent on the porosity and thus the permeability of the formation and may reach approximately 10 t/hr. Because the fluids used in petroleum technology are in some cases quite expensive, an extensive fluid loss may not be tolerable. Of course, there are also environmental reasons to prevent fluid loss.

MECHANISM OF ACTION OF FLUID LOSS AGENTS

Reduced fluid loss is achieved by plugging a porous rock in some way. The basic mechanisms are shown in Table 5.1.

Fluid Loss Measurement

The fluid loss process is complex and difficult to model because many of the parameters are not accessible for evaluation. Therefore, empirical models have been used, and some of them perform well. Also the protocols of testing influence the quality of the data. Some of the models have been reviewed and other models have been proposed that provide better fits to the available data (Clark, 2010).

Static fluid loss measurements provide inadequate results for comparing fracturing fluid materials or for understanding the complex mechanisms of viscous fluid invasion, filter cake formation, and filter cake erosion (Vitthal and McGowen, 1996b). On the other hand, dynamic fluid loss studies have inadequately addressed the development of proper laboratory methods, which has led to erroneous and conflicting results.

Results from a large-scale, high-temperature, high-pressure simulator were compared with laboratory data, and significant differences in spurt loss values were found (Lord et al., 1995).

Hydraulic Fracturing Chemicals and Fluids Technology. http://dx.doi.org/10.1016/B978-0-12-411491-3.00005-4
© 2013 Elsevier Inc. All rights reserved.

TABLE 5.1 Mechanisms of Fluid Loss Prevention

Particle Types	Description
Macroscopic	Suspended particles may clog the pores, forming a filter cake with reduced permeability
Microscopic	Macromolecules form a gel in the boundary layer of a porous formation
Chemical grouted	A resin is injected in the formation, which cures irreversibly; suitable for bigger caverns

Static experiments with piston-like filtering can be reliable, however, to obtain information on the fluid loss behavior in certain stages of a cementation process, in particular when the slurry is at rest.

Experimental studies indicated that two distinct stages of fluid loss can be differentiated (Vitthal and McGowen, 1996a):

1. An early high leakoff phase before a competent filter cake is established across the face of the formation (spurt loss).
2. A stage where all the fluid loss is controlled by the leakoff through the filter cake.

Action of Macroscopic Particles

A monograph concerning the mechanism of invasion of particles into the formation is given by Chin (1995).

One of the basic mechanisms in fluid loss prevention is shown in Figure 5.1. The fluid contains suspended particles. These particles move with the lateral flow out of the drill hole into the porous formation. The porous formation acts like a sieve for the suspended particles. The particles therefore will be captured near the surface and accumulated as a filter cake.

The hydrodynamic forces acting on the suspended colloids determine the rate of cake buildup and therefore the fluid loss rate. A simple model has been proposed in the literature that predicts a power law relationship between the filtration rate and the shear stress at the cake surface (Jiao and Sharma, 1994).

The model shows that the cake formed will be inhomogeneous with smaller particles being deposited as the filtration proceeds. An equilibrium cake thickness is achieved when no particles small enough to be deposited are available in the suspension. The cake thickness as a function of time can be computed from the model.

For a given suspension rheology and flow rate there is a critical permeability of the filter, below which no cake will be formed. The model also suggests that the equilibrium cake thickness can be precisely controlled by an appropriate choice of suspension flow rate and filter permeability.

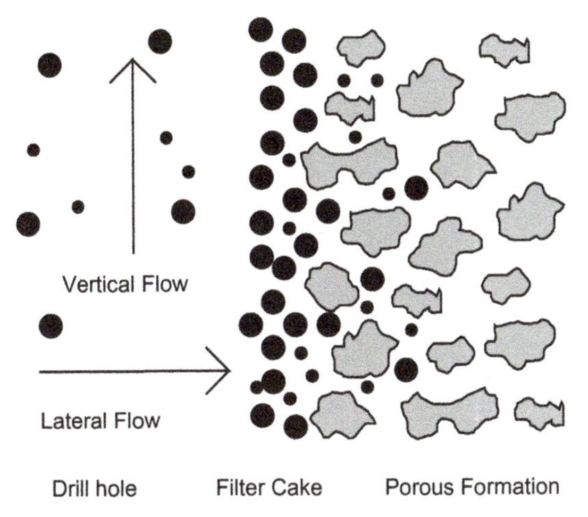

FIGURE 5.1 Formation of a filter cake in a porous formation from suspension (•) in a drilling fluid.

High-permeability fracturing zones can easily be damaged by deeply penetrating fluid leakoff along the fracture, or by the materials in the fluid to minimize the amount of leakoff. Several fracturing treatments in high-permeability formations, which are characterized by short lengths, and often by disproportionate widths, exhibit positive post-treatment skin effects. This is the result of fracture face damage (Aggour and Economides, 1996). If the invasion of the fracturing fluid is minimized, the degree of damage is of secondary importance.

So if the fluid leakoff penetration is small, even severe permeability impairments can be tolerated without exhibiting positive skin effects. The first priority in designing fracture treatments should be to maximize the conductivity of the fracture. In high-permeability fracturing, the use of high concentrations of polymer crosslinked fracturing fluids with fluid loss additives and breakers is recommended.

Materials used to minimize leakoff can also damage the conductivity of the proppant pack. High shear rates at the tip of the fracture may prevent the formation of external filter cakes, hence increasing the magnitude of spurt losses in highly permeable formations, so nondamaging additives are needed. Enzymatically degradable fluid loss additives are available. Table 5.2 summarizes some fluid loss additives suitable for hydraulic fracturing fluids.

The conductivity of fractures created with guar-based polymers may be low because of residual unbroken polymer gel remaining in the fracture. This residue can cause a permeability impairment in the proppant pack, resulting in a low fracture conductivity and decreased effective fracture length. An evaluation of two important aspects of the gel damage process has been performed, i.e., the thickness of the filter cake and the yield stress of the concentrated polymer gel that accumulates in the fracture (Xu et al., 2011).

TABLE 5.2 Fluid Loss Additives for Hydraulic Fracturing Fluids

Chemical	References
Calcium carbonate and lignosulfonate[a]	Johnson and Smejkal (1993) and Johnson (1996)
Natural starch	Elbel et al. (1995), Navarrete and Mitchell (1995), and Navarrete et al. (1996)
Carboxymethyl starch	Elbel et al. (1995), Navarrete and Mitchell (1995), and Navarrete et al. (1996)
Hydroxypropyl starch[b]	Elbel et al. (1995), Navarrete et al. (1996), and Navarrete and Mitchell (1995)
Hydroxyethyl cellulose (HEC) with crosslinked guar gums[c]	Cawiezel et al. (1999)
Granular starch and particulate mica	Cawiezel et al. (1999)

[a]*Wellan or xanthan gum polymer can be added to keep the calcium carbonate and lignosulfonate in suspension.*
[b]*Synergistic effect, see text.*
[c]*500 mD permeability.*

The thickness of the filter cake created during the leakoff process is a function of the polymer loading and the amount of leakoff. The thickness of the filter cake varies linearly with leakoff volume. This means that the gel concentration factor is constant for such a guar polymer fluid. The concentrated polymer filter cake created by leakoff exhibits a rheology-like Herschel-Bulkley fluids with a yield stress.

A Herschel-Bulkley fluid is a generalized non-Newtonian fluid. There, the strain experienced by the fluid is related to the stress in a rather complicated, nonlinear way, characterized by the consistency k, the flow index n, and the yield shear stress tensor τ_0. This fluid model was created in 1926 (Herschel and Bulkley, 1926).

The yield stress is a critical parameter that indicates whether the gel can be removed from the fracture. The yield stress of polymer filter cakes depends strongly on the concentration of both polymer and gel breaker (Xu et al., 2011).

ADDITIVE CHEMICALS

Granular Starch and Mica

A fluid loss additive has been described that consists of granular starch composition and fine particulate mica (Cawiezel et al., 1996). Mica belongs to the group of sheet silicates. Some mica minerals are listed in Table 5.3.

TABLE 5.3 Mica Minerals

Name	Formula
Biotite	$K(Mg,Fe)_3(AlSi_3)O_{10}(OH)_2$
Muscovite	$KAl_2(AlSi_3)O_{10}(OH)_2$
Phlogopite	$KMg_3(AlSi_3)O_{10}(OH)_2$
Lepidolite	$K(Li,Al)_{2--3}(AlSi_3)O_{10}(OH)_2$
Margarite	$CaAl2(Al2Si2)O_{10}(OH)2$
Glauconite	$(K,Na)(Al,Mg,Fe)_2(Si,Al)4O_{10}(OH)_2$

A method of fracturing a subterranean formation penetrated by a borehole comprises injecting into the borehole and into contact with the formation, at a rate and pressure sufficient to fracture the formation, a fracturing fluid containing the additive in an amount sufficient to provide fluid loss control.

Depolymerized Starch

Partially depolymerized starch provides decreased fluid losses at much lower viscosities than the corresponding starch derivatives that have not been partially depolymerized (Dobson and Mondshine, 1997).

Controlled Degradable Fluid Loss Additives

A fluid loss additive for a fracturing fluid has been described that comprises a mixture of natural and modified starches plus an enzyme (Williamson et al., 1991b). As stated already in detail before, the enzyme degrades the α-linkage of starch but does not degrade the β-linkage of guar and modified guar gums when used as a thickener.

Natural or modified starches are utilized in a preferred ratio of 3:7 to 7:3, with optimum at 1:1, and the mix is used in the dry form for application from the surface down the well. The preferred modified starches are the carboxymethyl and hydroxypropyl derivatives. Natural starches may be those of corn, potatoes, wheat, and soy, and the most preferred is corn starch.

Blends include two or more modified starches, as well as blends of natural and modified starches. Optionally, the starches are coated with a surfactant, such as sorbitan monooleate, ethoxylated butanol, or ethoxylated nonyl phenol, to aid dispersion into the fracturing fluid.

A fluid loss additive has been described (Williamson et al., 1991a) that helps achieve a desired fracture geometry by lowering the spurt loss and leakoff rate of the fracturing fluid into the surrounding formation by rapidly forming a filter cake with low permeability. The fluid loss additive is readily degraded after the completion of the fracturing process. The additive has a broad particulate size distribution that is ideal for use in effectively treating a wide range of formation porosities and is easily dispersed in the fracturing fluid.

This fluid loss additive is a blend of modified starches or blends of one or more modified starches and one or more natural starches. These blends have been found to maintain injected fluid within the created fracture more effectively than natural starches. The additive is subject to controlled degradation to soluble products by a naturally proceeding oxidation reaction or by bacterial attack by bacteria naturally present in the formation. The oxidation may be accelerated by adding oxidizing agents such as persulfates and peroxides.

Preferred peroxides are calcium peroxide and magnesium peroxide (Giffin, 2012). As persulfate breakers, ammonium persulfate, sodium persulfate, and potassium persulfate have been suggested (Card et al., 1999; Tulissi et al., 2012).

Degradation of Fluid Loss Additives

A fluid loss additive for fracturing fluids, which is a mixture of natural starch (corn starch) and chemically modified starches (carboxymethyl and hydroxypropyl derivatives) plus an enzyme, has been described (Williamson and Allenson, 1989; Williamson et al., 1991b). The enzyme degrades the α-linkage of starch but does not degrade the β-linkage of guar and modified guar gums when used as a thickener.

The starches can be coated with a surfactant, such as sorbitan monooleate, ethoxylated butanol, or ethoxylated nonyl phenol, to facilitate the dispersion into the fracturing fluid. Modified starches or blends of modified and natural starches with a broad particulate size distribution have been found to maintain the injected fluid within the created fracture more effectively than natural starches (Williamson et al., 1991a). The starches can be degraded by oxidation or by bacterial attack.

Succinoglycan

Succinoglycan is a biopolymer. It has been shown to possess a combination of desirable properties for fluid loss control (Lau (1994a,1994b)). These include ease of mixing, cleanliness, shear thinning rheology, temperature-insensitive viscosity below its transition temperature (T_m), and an adjustable transition temperature (T_m) over a wide range of temperatures. Succinoglycan fluids rely solely on viscosity to reduce fluid loss. It does not form a hard-to-remove filter cake, which can cause considerable formation damage.

Based on these findings, succinoglycan has been used successfully as a fluid loss pill before and after gravel packing in more than 100 offshore wells. Calculations based on laboratory-measured rheology and field experience have shown succinoglycan to be effective in situations in which HEC is not. Fluid loss, even over 40 barrels/h, was reduced to several barrels per hour after application of a properly designed succinoglycan pill. Most wells experienced no problem in production after completion.

Succinoglycan can be degraded with an internal acid breaker (Bouts et al., 1997). The formation damage that results from the incomplete back-production of viscous fluid loss control pills can be minimized if a slow-acting internal breaker is employed. In particular, core-flow tests have indicated that combining a succinoglycan-based pill with a hydrochloric acid internal breaker enables a fluid loss system with sustained control followed by delayed break back and creates only low levels of impairment. To describe the delayed breaking of the succinoglycan/hydrochloric acid system, a model, based on bond breaking rate, has been used.

With this model, it is possible to predict the change of the rheological properties of the polymer as a function of time for various formation temperatures, transition temperatures of the succinoglycan, and acid concentrations. The model can be used to identify optimal formulations of succinoglycan and acid breaker on the basis of field requirements, such as the time interval over which fluid loss control is needed, the overbalance pressure a pill that should be able to withstand, and the brine density required.

Scleroglucan

A combination of graded calcium carbonate particle sizes, a nonionic polysaccharide of the scleroglucan type, and a modified starch, has been claimed for use in fluid loss formulations (Johnson, 1996). It is important that the calcium carbonate particles are distributed across a wide size range to prevent filtration or fluid loss into the formation. Because the filter cake particles do not invade the wellbore due to the action of the biopolymer and the starch, no high-pressure spike occurs during the removal of the filter cake.

The rheological properties of the fluid allow it to be used in a number of applications in which protection of the original permeable formation is desirable. The applications include drilling, fracturing, and controlling fluid losses during completion operations, such as gravel packing or well workovers.

Poly(orthoester)s

Aliphatic poly(ester)s degrade chemically by a hydrolytic cleavage. The process of hydrolysis can be catalyzed by both acids and bases. During the hydrolysis, carboxylic end groups are formed during chain scission, and this may enhance the rate of further hydrolysis. This mechanism is known in the art as autocatalysis and is thought to make poly(ester) matrices more bulk-eroding.

Among the esters, poly(orthoester)s and aliphatic poly(ester)s, i.e., poly (lactide)s, are preferred. Poly(lactide)s can be synthesized either from lactic acid by a condensation reaction or more commonly by ring opening polymerization of a cyclic lactide monomer.

TABLE 5.4 Fracturing Fluid Composition (Todd et al., 2006)

Compound	Tradename[a]	%
Water		
Potassium chloride		1
De-emulsifier	LO-SURF 300	0.05
Poly(ester)		0.15
Guar		0.2
Buffer (CH_3COOH)	BA-20	0.005
Caustic	MO-67	0.1
Borate Crosslinking agent	CL-28M	0.05
Gel breaker ($NaClO_3$)	VICON NF	0.1
Bactericide (2,2-dibromo-3-nitrilopropionamide)	BE-3S	0.001
Bactericide (2-Bromo-2-nitro-1,3-propanediol)	BE-6	0.001
Fracturing sand		50

[a]*Halliburton Energy Services, Inc.*

The degradation by hydrolysis should take place rather slowly over time. In fracturing operations, the material should not begin to degrade until after the proppant has been placed in the fracture. Moreover, the slow degradation helps to provide fluid loss control during the placement of the proppant (Todd et al., 2006). An example of a fracturing fluid composition is given in Table 5.4.

Poly(hydroxyacetic acid)

A low-molecular-weight condensation product of hydroxyacetic acid with itself or compounds containing other hydroxy acid, carboxylic acid, or hydroxy carboxylic acid moieties has been suggested as a fluid loss additive (Bellis and McBride, 1987). Production methods of the polymer have been described.

The reaction products are ground to 0.1–1500 μm particle size. The condensation product can be used as a fluid loss material in a hydraulic fracturing process in which the fracturing fluid comprises a hydrolyzable, aqueous gel.

The hydroxyacetic acid condensation product hydrolyzes at formation conditions to provide hydroxyacetic acid, which breaks the aqueous gel autocatalytically and eventually provides the restored formation permeability without the need of the separate addition of a gel breaker (Cantu et al. (1990,1989); Casad et al., 1991).

Polyphenolics

Organophilic polyphenolic materials for oil-based drilling muds have been described (Cowan et al., 1988). The additives are prepared from a polyphenolic material and one or more phosphatides. The phosphatides are phosphoglycerides obtained from vegetable oils, preferably commercial lecithin.

Humic acids, lignosulfonic acid, lignins, phenolic condensates, tannins; the oxidized, sulfonated, or sulfomethylated derivatives of these polyphenolic materials may serve as polyphenolic materials.

A fluid loss additive is described that uses graded calcium carbonate particle sizes and a modified lignosulfonate (Johnson and Smejkal, 1993). Optionally, a thixotropic polymer, such as a wellan or xanthan gum polymer, is used to keep the $CaCO_3$ and lignosulfonate in suspension.

It is important that the calcium carbonate particles are distributed across a wide size range to prevent filtration or fluid loss into the formation. Furthermore, the lignosulfonate must be polymerized to an extent effective to reduce its water solubility. The modified lignosulfonate is necessary for the formation of a filter cake essentially at the surface of the wellbore.

Because the filter cake particles do not invade the wellbore due to the action of the modified lignosulfonate, no high-pressure spike occurs during the removal of the filter cake, which would indicate damage of the formation and wellbore surface. The additive is useful in fracturing fluids, completion fluids, and workover fluids.

Tests showed that a fluid loss additive on the base of a sulfonated tannic-phenolic resin is effective for fluid loss control at high temperature and pressure, and it exhibits a good resistance to salt and acid (Huang, 1996).

Phthalimide as a Diverting Material

Phthalimide, c.f., Figure 5.2, has been described as a diverting material, or fluid loss additive, for diverting aqueous treating fluids, including acids, into progressively less permeable portions of a subterranean formation (Dill, 1987).

The additive also reduces the fluid loss to the formation of an aqueous or hydrocarbon treating fluid utilized, for example, in fracturing treatments. The performance of the material depends on the particle size of the material.

Phthalimide will withstand high formation temperatures and can be readily removed from the formation by dissolution in the produced fluids or by sublimation at elevated temperatures. The material is compatible both with other formation permeability-reducing materials and formation permeability-increasing materials. The phthalimide particles act by sealing off portions of a

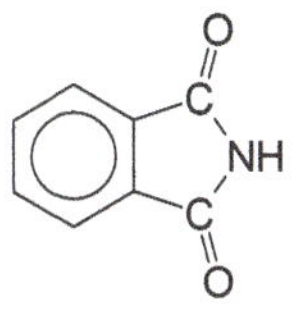

O
C
NH
C
O

FIGURE 5.2 Phthalimide.

subterranean formation by blocking off the fissures, pores, channels, and vugs that grant access to the formation from the wellbore penetrating the formation.

Viscoelastic Additives

Reagents that generate the viscoelasticity in the surfactant solutions are salts, e.g., ammonium chloride, potassium chloride, sodium salicylate, and sodium isocyanate. In addition, nonionic organic molecules, such as chloroform, are active in generating viscoelasticity. The electrolyte content of surfactant solutions is also an important parameter for the viscoelasticity (Sullivan et al., 2006). Aqueous fluids gelled with viscoelastic surfactants (VES)s have been used for hydraulic fracturing operations.

However, the same property that makes VES fluids less damaging tends to result in significantly higher fluid leakage into the reservoir matrix, which reduces the efficiency of the fluid especially during VES fracturing treatments. Thus, it is important to use fluid loss agents for VES fracturing treatments in high-permeability formations (Huang and Crews, 2009).

The fluid loss properties of such fluids can be improved by the addition of mineral oil which has a viscosity >20 MPa s at ambient temperature. The mineral oil may initially be dispersed as oil droplets in an internal, discontinuous phase of the fluid. The mineral oil is added to the fluid after it has been substantially gelled.

In the experiments reproduced below, tallowamidopropylamine oxide has been used as VES surfactant, which is available from Akzo Nobel (Podwysocki, 2004). It has been demonstrated that the viscosity of aqueous fluids containing 3% KCl and gelled with 6% VES at 66 °C without and with 2% mineral oil is adversely affected. This behavior is in contrast to other observations, since larger amounts of hydrocarbons and mineral oils tend to inhibit or break the gel of VES-gelled fluids (Huang and Crews, 2009).

On the other hand, the fluid loss is improved as shown in model experiments. The fluid loss as a function of testing time is shown in Figure 5.3. The test is conducted at 0.7 MPa with 400 mD ceramic disks at 66 °C.

It has been discovered that the addition of magnesium oxide, or calcium hydroxide, to an aqueous fluid gelled with a VES improves the fluid loss of these brines (Huang et al., 2009).

It is of importance that these fluid loss control agents dissolve slowly, which permits their easy removal from the formation and complete with little or no damage to the formation.

The introduction of these additives to a VES-gelled aqueous system will limit and reduce the amount of VES fluid which leaksoff into the pores of a reservoir during a fracturing or frac-packing treatment, thus minimizing the formation damage that may occur by the VES fluid within the reservoir pores.

Moreover, the range in reservoir permeability does not significantly control the rate of fluid loss. Thus, the rate of leakoff in 2000 mD reservoirs will be

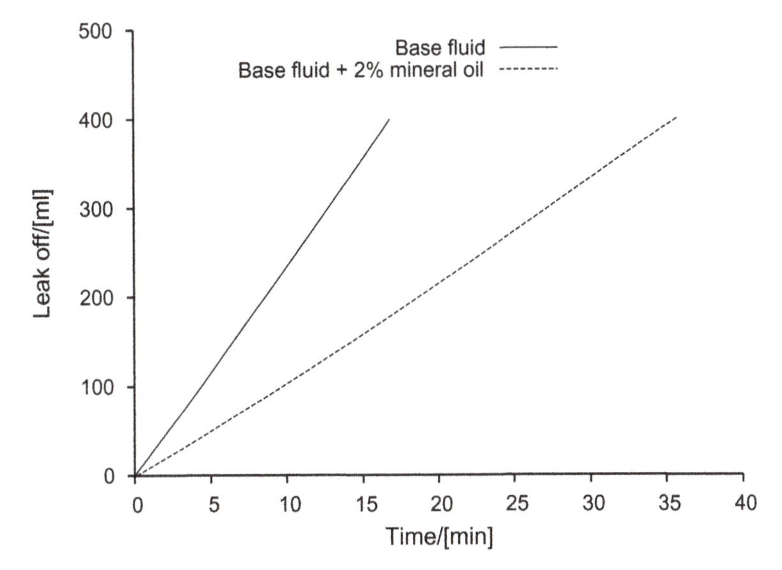

FIGURE 5.3 Fluid loss versus time (Huang and Crews, 2009).

comparable to the rate of leakoff in 100 mD reservoirs. This behavior expands the range in reservoir permeability to which the VES fluid may be applied.

It is believed that the fluid loss agents associate with the VES micelles. As the VES fluid is leaked off into the reservoir, a viscous layer of micelles and fluid loss control particles accumulate on the formation face, thus reducing the rate of VES fluid leakoff.

Particulate plugging of the reservoir pores is not the mechanism of leakoff control. Tests with nanometer-sized fluid loss agents that definitely cannot bridge or plug reservoir pores of 1 mD or higher reservoir permeability still develop a viscous micelle layer. Thus, the size of the fluid loss agent is not a controlling or primary factor for controlling the leakoff rate (Huang et al., 2009).

Degradation of Fluid Loss Additives

A fluid loss additive for fracturing fluids, which is a mixture of natural starch (corn starch) and chemically modified starches (carboxymethyl and hydroxypropyl derivatives) plus an enzyme, has been described (Williamson and Allenson, 1989; Williamson et al., 1991b). The enzyme degrades the α-linkage of starch but does not degrade the β-linkage of guar and modified guar gums when used as a thickener.

The starches can be coated with a surfactant, such as sorbitan monooleate, ethoxylated butanol, or ethoxylated nonyl phenol, to facilitate the dispersion into the fracturing fluid. Modified starches or blends of modified and natural starches with a broad particulate size distribution have been found to maintain the injected fluid within the created fracture more effectively than natural starches (Williamson et al., 1991a). The starches can be degraded by oxidation or by bacterial attack.

Tradenames in References

Tradename	Supplier
Description	
Britolo® 35 USP	Sonneborn Refined Products
High-viscosity mineral oil (Huang and Crews, 2009)	
ClearFRAC™	Schlumberger Technology Corp.
Stimulating Fluid (Huang and Crews, 2009; Huang et al., 2009)	
Diamond FRAQ™	Baker Oil Tools
VES breaker (Huang and Crews, 2009)	
Gloria®	Sonneborn Refined Products
High-viscosity mineral oil (Huang and Crews, 2009)	
Hydrobrite® 200	Sonneborn Inc.
White mineral oil (Huang and Crews, 2009)	
Kaydol® oil	Witco Corp.
Mineral oil (Huang and Crews, 2009)	
Performance® 225 N	ConocoPhillips
Base Oil (Huang and Crews, 2009)	
SurFRAQ™ VES	Baker Oil Tools
Tallowamidopropylamine oxide (Huang and Crews, 2009; Huang et al., 2009)	

REFERENCES

Aggour, T.M., Economides, M.J., 1996. Impact of fluid selection on high-permeability fracturing. In: Proceedings Volume. vol. 2. SPE Europe Petroleum Conference, Milan, Italy, 22–24 October 1996, pp. 281–287.

Bellis, H.E., McBride, E.F., 1987. Composition and method for temporarily reducing the permeability of subterranean formations. EP Patent 228 196, assigned to Du Pont De Nemours & Co., 8 July 1987.

Bouts, M.N., Trompert, R.A., Samuel, A.J., 1997. Time delayed and low-impairment fluid-loss control using a succinoglycan biopolymer with an internal acid breaker. SPE J. 2 (4), 417–426.

Cantu, L.A., McBride, E.F., Osborne, M.W., 1989. Formation fracturing process. US Patent 4 848 467, assigned to Conoco Inc. and Du Pont De Nemours & Co., 18 July 1989.

Cantu, L.A., McBride, E.F., Osborne, M., 1990. Well treatment process. EP Patent 404 489, assigned to Conoco Inc. and Du Pont De Nemours & Co., 27 December 1990.

Card, R.J., Nimerick, K.H., Maberry, L.J., McConnell, S.B., Nelson, E.B., 1999. On-the-fly control of delayed borate-crosslinking of fracturing fluids. US Patent 5 877 127, assigned to Schlumberger Technology Corporation, Sugar Land, TX, 2 March 1999. <http://www.freepatentsonline.com/5877127.html>.

Casad, B.M., Clark, C.R., Cantu, L.A., Cords, D.P., McBride, E.F., 1991. Process for the preparation of fluid loss additive and gel breaker. US Patent 4 986 355, assigned to Conoco Inc., 22 January 1991.

Cawiezel, K.E., Navarrete, R., Constien, V., 1996. Fluid loss control. GB Patent 2 291 906, assigned to Sofitech NV, 7 February 1996.

Cawiezel, K.E., Navarrete, R.C., Constien, V.G., 1999. Fluid loss control. US Patent 5 948 733, assigned to Dowell Schlumberger Inc., 7 September 1999.

Chin, W.C., 1995. Formation Invasion: With Applications to Measurement-While-Drilling, Time Lapse Analysis, and Formation Damage. Gulf Publishing Co., Houston.

Clark, P.E., 2010. Analysis of fluid loss data II: models for dynamic fluid loss. J. Petrol. Sci. Eng. 70 (3–4), 191–197. http://dx.doi.org/10.1016/j.petrol.2009.11.010.

Cowan, J.C., Granquist, V.M., House, R.F., 1988. Organophilic polyphenolic acid adducts. US Patent 4 737 295, assigned to Venture Chemicals Inc., 12 April 1988.

Dill, W.R., 1987. Diverting material and method of use for well treatment. CA Patent 1 217 320, 3 February 1987.

Dobson, J.W., Mondshine, K.B., 1997. Method of reducing fluid loss of well drilling and servicing fluids. EP Patent 758 011, assigned to Texas United Chem. Co. Llc., 12 February 1997.

Elbel, J.L., Navarrete, R.C., Poe Jr., B.D., 1995. Production effects of fluid loss in fracturing high-permeability formations. In: Proceedings Volume. SPE Europe Formation Damage Control Conference, The Hague, Netherlands, 15–16 May 1995, pp. 201–211.

Giffin, W.J., 2012. Compositions and processes for fracturing subterranean formations. US Patent 8 293 687, assigned to Titan Global Oil Services Inc., Bloomfield Hills, MI, 23 October 2012. <http://www.freepatentsonline.com/8293687.html>.

Harris, P., 1988. Fracturing-fluid additives. J. Petrol. Technol. 40 (10). http://dx.doi.org/10.2118/17112-PA.

Herschel, W.H., Bulkley, R., 1926. Konsistenzmessungen von Gummi-Benzollösungen. Kolloid-Zeitschrift 39 (4), 291–300. http://dx.doi.org/10.1007/BF01432034.

Huang, N., 1996. Synthesis of fluid loss additive of sulfonate tannic-phenolic resin. Oil Drilling Prod. Technol. 18 (2), 39–42, 106–107.

Huang, T., Crews, J.B., 2009. Use of mineral oils to reduce fluid loss for viscoelastic surfactant gelled fluids. US Patent 7 615 517, assigned to Baker Hughes Inc., Houston, TX, 10 November 2009. <http://www.freepatentsonline.com/7615517.html>.

Huang, T., Crews, J.B., Treadway Jr., J.H., 2009. Fluid loss control agents for viscoelastic surfactant fluids. US Patent 7550413, assigned to Baker Hughes Inc., Houston, TX, 23 June 2009. <http://www.freepatentsonline.com/7550413.html>.

Jiao, D., Sharma, M.M., 1994. Mechanism of cake buildup in crossflow filtration of colloidal suspensions. J. Colloid Interface Sci. 162 (2), 454–462.

Johnson, M., 1996. Fluid systems for controlling fluid losses during hydrocarbon recovery operations. EP Patent 691 454, assigned to Baker Hughes Inc., 10 January 1996.

Johnson, M.H., Smejkal, K.D., 1993. Fluid system for controlling fluid losses during hydrocarbon recovery operations. US Patent 5 228 524, assigned to Baker Hughes Inc., 20 July 1993.

Lau, H.C., 1994a. Laboratory development and field testing of succinoglycan as a fluid-loss-control fluid. SPE Drill. Completion 9 (4), 221–226.

Lau, H.C., 1994b. Laboratory development and field testing of succinoglycan as fluid-loss-control fluid. SPE Peer Approved Paper.

Lord, D.L., Vinod, P.S., Shah, S., Bishop, M.L., 1995. An investigation of fluid leakoff phenomena employing a high-pressure simulator. In: Proceedings Volume. Annual SPE Technical Conference, Dallas, 22–25 October 1995, pp. 465–474.

Navarrete, R.C., Mitchell, J.P., 1995. Fluid-loss control for high-permeability rocks in hydraulic fracturing under realistic shear conditions. In: Proceedings Volume. SPE Production and Operations Symposium, Oklahoma, City, 2–4 April 1995, pp. 579–591.

Navarrete, R.C., Brown, J.E., Marcinew, R.P., 1996. Application of new bridging technology and particulate chemistry for fluid-loss control during fracturing highly permeable formations. In: Proceedings Volume. vol. 2. SPE Europe Petroleum Conference, Milan, Italy, 22–24 October 1996, pp. 321–325.

Podwysocki, M., 2004. Akzo nobel surfactants. Technical Bulletin SC05-0707, Akzo Nobel Surface Chemistry LLC, 525 W. Van Buren Street Chicago, IL 60607-3823.

Rimassa, S., Howard, P., Blow, K., 2009. Optimizing fracturing fluids from flowback water. In: Proceedings of SPE Tight Gas Completions Conference. No. 125336-MS. SPE Tight Gas Completions Conference, 15–17 June 2009, San Antonio, Texas, USA, Society of Petroleum Engineers, Dallas, Texas, pp. 1–8. http://dx.doi.org/10.2118/125336-MS.

Sullivan, P., Christanti, Y., Couillet, I., Davies, S., Hughes, T., Wilson, A., 2006. Methods for controlling the fluid loss properties of viscoelastic surfactant based fluids. US Patent 7081439, assigned to Schlumberger Technology Corporation, Sugar Land, TX, 25 July 2006. <http://www.freepatentsonline.com/7081439.html>.

Todd, B.L., Slabaugh, B.F., Munoz Jr., T., Parker, M.A., 2006. Fluid loss control additives for use in fracturing subterranean formations. US Patent 7096947, assigned to Halliburton Energy Services, Inc., Duncan, OK, 29 August 2006. http://www.freepatentsonline.com/7096947.html.

Tulissi, M.G., Luk, S., Vaughan, J., Browne, D.J., Dusterhoft, D., 2012. Fracturing method and apparatus utilizing gelled isolation fluid. US Patent 8141638, assigned to Trican Well Services Ltd., Calgary, CA, 27 March 2012. <http://www.freepatentsonline.com/8141638.html>.

Vitthal, S., McGowen, J., 1996a. Fracturing fluid leakoff under dynamic conditions: Part 2: Effect of shear rate, permeability, and pressure. In: Proceedings of SPE Annual Technical Conference and Exhibition. No. 36493-MS. SPE Annual Technical Conference and Exhibition, 6–9 October 1996, Denver, Colorado, Society of Petroleum Engineers, pp. 821–835. http://dx.doi.org/10.2118/36493-MS.

Vitthal, S., McGowen, J.M., 1996b. Fracturing fluid leakoff under dynamic conditions: Part.2: Effect of shear rate, permeability, and pressure. In: Proceedings Volume. Annual SPE Technical Conference, Denver, 6–9 October 1996, pp. 821–835.

Williamson, C.D., Allenson, S.J., 1989. A new nondamaging particulate fluid-loss additive. In: Proceedings Volume. SPE International Symposium on Oilfield Chemistry, Houston, 8–10 February 1989, pp. 147–158.

Williamson, C.D., Allenson, S.J., Gabel, R.K., 1991a. Additive and method for temporarily reducing permeability of subterranean formations. US Patent 4997581, assigned to Nalco Chemical Co., 5 March 1991.

Williamson, C.D., Allenson, S.J., Gabel, R.K., Huddleston, D.A., 1991b. Enzymatically degradable fluid loss additive. US Patent 5032297, assigned to Nalco Chemical Co., 16 July 1991.

Xu, B., Hill, A.D., Zhu, D., Wang, L., 2011. Experimental evaluation of guar-fracture-fluid filter-cake behavior. SPE Prod. Oper. 26 (4), 381–387.

Chapter 6

Emulsifiers

Emulsions play an important role in fluids used for oilfield applications. These include most importantly drilling and treatment fluids. Actually in these fields of applications, the emulsions are not addressed as such, rather, e.g., as oil-based drilling muds or water-based drilling muds, which are essentially emulsions from the view of physics. Here we summarize some basic aspects of emulsions.

Oilfield emulsions are sometimes classified based on their degree of kinetic stability (Kokal and Wingrove, 2000; Kokal, 2006):

- *Loose emulsions:* Those that will separate within a few minutes. The separated water is sometimes referred to as the free water.
- *Medium emulsions:* They will separate in some 10 min.
- *Tight emulsions:* They will separate within hours, days, or even weeks. Thereafter, the separation may not be complete.

Emulsions are also classified by the size of the droplets in the continuous phase. When the dispersed droplets are larger than 0.1 μm, the emulsion is a macroemulsion (Kokal, 2006).

From the purely thermodynamic view, an emulsion is an unstable system. This arises because there is a natural tendency for a liquid-liquid system to separate and reduce its interfacial area and thus its interfacial energy (Kokal and Wingrove, 2000).

A second class of emulsions is known as microemulsion. Such emulsions are formed spontaneously when two immiscible phases are brought together with extremely low interfacial energy. Microemulsions have very small droplet sizes, less than 10 nm, and are stable from the view of thermodynamics. Microemulsions are fundamentally different from macroemulsions in their formation and stability. The use of microemulsion systems in oil industry has been reviewed (Santanna et al., 2012).

OIL-IN-WATER EMULSIONS

Oil-in-water emulsions have also been proposed for use in fracturing operations. In order to provide the emulsion with sufficiently high viscosities, the concentration of the oil phase in these systems must be extremely high, in

Hydraulic Fracturing Chemicals and Fluids Technology. http://dx.doi.org/10.1016/B978-0-12-411491-3.00006-6
© 2013 Elsevier Inc. All rights reserved.

the order of 95%. At these high concentrations, a considerable distortion of the discrete oil particles from the usual spherical shape occurs. Such systems are difficult to handle and generally exhibit high friction loss in the well conduit. Moreover, the emulsions are difficult to stabilize. Additives such as film strengtheners, inorganic salts, and deliquescent salts have been proposed for improving the properties of the oil-in-water emulsions (Kiel, 1973). These compositions are more favorable than those previously developed.

Thickening agents which possess surface active properties can be used as an emulsifier to promote the oil-in-water emulsion. For example, poly(vinylcarboxylic acid) neutralized with a long-chain amine and a common base such as sodium hydroxide is capable of promoting extremely stable oil-in-water emulsions. The emulsified system exhibits excellent temperature resistance and therefore can be used in deep, high-temperature wells (Kiel, 1973).

INVERT EMULSIONS

Invert emulsions have a continuous phase that is an oleaginous fluid, and a discontinuous phase that is a fluid, which is at least partially immiscible in the oleaginous fluid. Actually, invert emulsions are water-in-oil emulsions.

Invert emulsions may have desirable suspension properties for particulates like drill cuttings. As such, they can easily be weighted if desired. It is well known to reverse invert emulsions to regular emulsions by changing the pH or by protonating the surfactant. In this way, the affinity of the surfactant for the continuous and discontinuous phases is changed (Taylor et al., 2009).

For example, if a residual amount of an invert emulsion remains in a wellbore, this portion may be reversed to a regular emulsion to clean out the emulsion from the wellbore. Invert emulsion compositions can be used, where the organic phase is gelled. For example, diesel can be gelled with decanephosphonic acid monoethyl ester and a Fe^{3+} activator (Taylor et al., 2009).

Polymers are often used to increase the viscosity of an aqueous fluid. The polymer should interact with this fluid as it should show a tendency to hydrate. Microemulsions may be helpful to achieve this target (Jones and Wentzler, 2008).

WATER-IN-WATER EMULSIONS

When two or more different water-soluble polymers are dissolved together in an aqueous medium, it is sometimes observed that the system phase separates into distinct regions. Particularly, this happens when two polymers are chosen that are each water soluble but thermodynamically incompatible with each other.

These emulsions are termed as water-in-water emulsions, or also aqueous two-phase systems (Sullivan et al., 2010).

In the food industry, such fluids are used to create polymer solutions that mimic the properties of fat globules. In the biomedical industry, such systems are exploited as separation media for proteins, enzymes, and other macromolecules that preferentially partition to one polymer phase in the mixture.

Aqueous two-phase polymer-polymer systems are also of interest in oilfield applications. These mixtures can be used to create low-viscosity pre-hydrated concentrated mixtures to allow the rapid mixing of the polymers at a well site to achieve a low viscosity polymer fluid.

Solutions of guar and hydroxypropyl cellulose (HPC) form aqueous phase-separated solutions over a range of polymer concentrations. A phase-separated mixture can be formed by simultaneously dissolving dry guar and dry HPC in a blender. After continued stirring the solution is allowed to rest to achieve phase separation. Afterwards, the phase-separated solution can be gently stirred to remix the guar-rich and HPC-rich phases. The two-phase polymer solution can be activated to become an elastic gel by thermal treatment, or by changing the ionic strength (Sullivan et al., 2010). This behavior can be used for zone isolation.

OIL-IN-WATER-IN-OIL EMULSIONS

Oil-in-water-in-oil emulsions can be used as a drive fluid for enhanced oil recovery operations or as a lubrication fluid. Such emulsions exhibit improved shear stability and shear thinning characteristics in comparison to water-in-oil emulsions.

An oil-in-water-in-oil emulsion is prepared from an oil-in-water emulsion that is subsequently dispersed in a second oil (Varadaraj, 2010). The second oil may contain a stabilizer, i.e., micron to sub-micron sized solid particles, naphthenic acids, and asphaltenes.

MICROEMULSIONS

A microemulsion is a thermodynamically stable fluid. It is different from kinetically stable emulsions which will be break into oil and water over time. The particle size of microemulsions ranges from about 10 nm to 300 nm. Because of the small particle sizes, microemulsions appear as clear or translucent solutions.

The particle sizes of microemulsions may be identified by dynamic light scattering or neutron scattering. Microemulsions have ultralow interfacial tension between the water phase and the oil phase.

Water-in-oil microemulsions have been known to deliver water-soluble oilfield chemicals into subterranean rock formations. Also known are

TABLE 6.1 Microemulsion with Corrosion Inhibitor (Yang and Jovancicevic, 2009)

Component	Amount (%)
Toluene	2
Oleic imidazoline (corrosion inhibitor)	4
Oleic acid (corrosion inhibitor)	4
Dodecyl benzene sulfonic acid	2
Ethanolamine	2
Butyl alcohol	20
Water	66

oil-in-alcohol microemulsions containing corrosion inhibitors in antifreeze compositions (Yang and Jovancicevic, 2009).

Microemulsions can be used to deliver a wide variety of oil-soluble oilfield chemicals, including corrosion inhibitors, asphaltene inhibitors, scale inhibitors, etc. Thereby the amount of organic solvent needed is reduced. The microemulsion increases the dispersibility of oilfield chemicals into the produced fluids or pumped fluids. In this way, the performance of the particular chemical is increased (Yang and Jovancicevic, 2009).

Microemulsions may be broken by a variety of mechanisms, such as by chemicals or by temperature changes. However, the most simple way of breaking is by dilution.

An example of a microemulsion that carries a corrosion inhibitor is listed in Table 6.1. The formulation in Table 6.1 can be easily diluted to the water phase. If the amount of toluene is increased in favor of water, a microemulsion is obtained that can be diluted by a hydrocarbon solvent. Still other examples are given elsewhere (Yang and Jovancicevic, 2009).

SOLIDS-STABILIZED EMULSION

Emulsions can be stabilized using undissolved solid particles, which are partially oleophilic (Bragg, 2000). Three-phase emulsions that are stabilized with solids have been reviewed (Menon and Wasan, 1988). Further, the phenomenon of oil loss due to entrainment in emulsion sludge layers has been assessed. A semiempirical approach has suggested for estimating the loss of oil.

The solid particles may be either indigenous to the formation or obtained from outside the formation. Nonformation solid particles include clays, quartz, feldspar, gypsum, coal dust, asphaltenes, and polymers. Preferably, however, the particles contain small amounts of a ionic compound. Typically, the particles exhibit some composite irregular shape (Bragg, 2000).

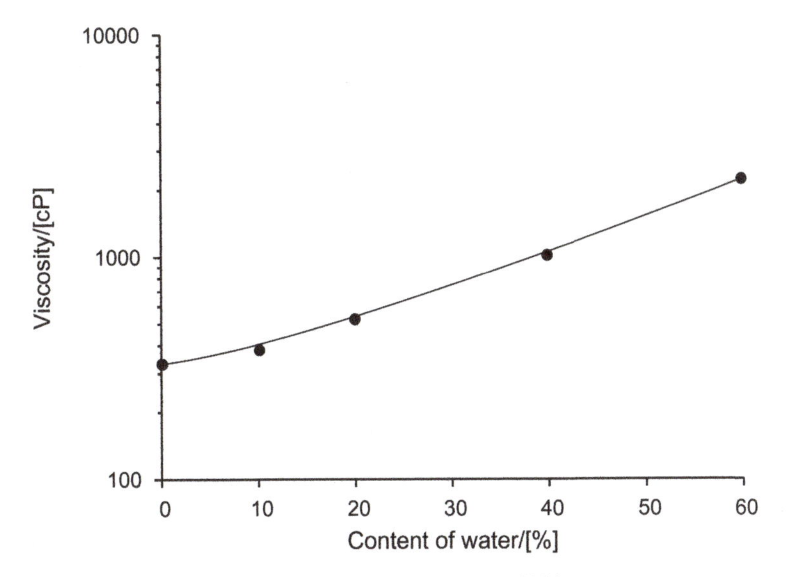

FIGURE 6.1 Viscosity versus water content (Varadaraj et al., 2004).

The solid particles should have either some oleophilic character for making an oil-external emulsion or some hydrophilic character for making a water-external emulsion. This property is important for ensuring that the particles can be wetted by the external continuous phase that holds the internal, discontinuous phase.

The oleophilic or hydrophilic character may be an inherent characteristic of the solid particles or else acquired by a chemical treatment of the particles.

For example, oleophilic fumed silicas, e.g., Aerosil™ R972 or CAB-O-SIL™, consist of small spheres of fumed silica that have been treated with organosilanes or organosilazanes to make the surfaces oleophilic. These are effective solids for stabilizing many crude oil emulsions. Such particles are extremely small, having primary particles consisting of spheres with diameters as small as about 10–20 nm, although the primary particles interact to form larger aggregates. These silicas are effective at concentrations of 0.5–20 $g\,l^{-1}$.

Figure 6.1 shows the viscosity of an emulsion with solid particles at a shear rate of 75 s^{-1} as a function of the water content.

The oil in the emulsion can be pretreated with a sulfonating agent prior to emulsification, in order to enhance its ability to make a solids-stabilized water-in-oil emulsion (Varadaraj et al., 2004). A specific procedure for the sulfonation of both the solids and the oil has been disclosed (Varadaraj et al., 2004):

Preparation 6.1. The crude oil and the solid particles are co-sulfonated. Twelve grams of crude oil and solids comprised of 0.06 g of 2-methylbenzyl tallow intercalated monomorillonite and 0.12 g of asphalt are stirred at 50 °C for 72 h. Subsequently, concentrated sulfuric acid is added at 3 parts per 100 parts of oil at 50 °C during 24 h. ■

TABLE 6.2 Effect of Various Methods of Pretreatment of Crude Oil (Varadaraj et al., 2007)

Method	Saturated	Unsaturated	NSO[a]	Asphaltene
		Compounds (%)		
Untreated crude oil	35.4	39.8	15.4	9.4
Dye sensitized photochemically treated	34.2	26.6	26.6	12.7
Photochemically treated	31.1	20.5	30.7	17.9
Thermally air oxidized	34.2	19.3	33.6	13.0
Biologically oxidized	32.4	39.8	18.4	9.4

[a]*Polycondensed aromatic benzene units with oxygen, nitrogen, and sulfur (NSO compounds).*

In a variant of the Preparation 6.1, the sulfonated crude oil can be combined with a maleic acid grafted copolymer from ethylene and propylene (Varadaraj et al., 2007).

In addition, instead of sulfonation, a photochemical treatment of the oil has been examined (Varadaraj et al., 2007). Hereby, a bentonite clay gel is mixed with the crude oil before photochemical treatment. In a dye sensitized photochemical treatment process, the crude oil is first mixed with Rhodamine-B. Rhodamine-B is a red dye that increases the quantum efficiency of the photochemical conversion of oil into oxidized products. In Table 6.2, the differences in change of the compositions of a crude oil by various methods of pretreatment are shown.

BIOTREATED EMULSION

Another pretreatment method to enhance the stability of a water-in-oil emulsion consists in biotreating the oil prior to emulsification. Water-in-oil emulsions made from a biotreated oil exhibit an enhanced stability. Oil degrading microbes are used in the biotreatment process (Varadaraj et al., 2007).

Preparation 6.2. For the biotreatment of the oil, the oil is placed into a bioreactor. Water-in a 10–100-fold excess should be present in the reactor. Oil degrading microbes, e.g., inoculum, are then added to the reactor. Inoculum is a culture of microbes. The concentration of microbes in the inoculum can be measured by colony-forming units. The oil degrading microbes can be obtained from an oil wastewater treatment facility.

Nutrients can be provided to feed the microbes. The nutrients will preferably contain nitrogen and phosphorus. The bioreactor is purged with air or oxygen at a temperature of 20–70 °C.

After the biotreatment step, the biotreated oil can be separated from the aqueous phase of the bioreaction prior to forming a water-in-oil emulsion with the biotreated oil. However, it is preferred to form an emulsion using both the biotreated oil and the aqueous phase of the bioreaction as the aqueous phase contains components that will help further enhance the stability of the resulting water-in-oil emulsion. ■

It is believed that the biotreatment step promotes an enhanced stability of a water-in-oil emulsion according to the following mechanisms (Varadaraj et al., 2007):

- Some of the aliphatic components of oil are oxidized and polar ketones or acid functionality are introduced on the aliphatic chain. Organo sulfur compounds are also susceptible to oxidization and can form corresponding sulfoxides. The oxygenated compounds are more surface active than the aliphatic components themselves and thus contribute to improving the stability of the water-in-oil emulsion.
- If naphthenic acids are present as salts of divalent cations like calcium, biooxidation is likely to convert these salts to decarboxylated naphthenic hydrocarbons or lower carbon number naphthenic acids and the corresponding metal oxide. These constituents serve to enhance the stability of the water-in-oil emulsion.
- In the process of biotreating an oil, the aqueous phase of the bioreaction also undergoes substantial changes. Upon completion of the bioreaction, the aqueous phase is a dispersion of biosurfactants, i.e., rhamnolipids produced by the microbes, and dead microbe cells. These components act synergistically to enhance the stability of water-in-oil emulsions. The aqueous phase of the bioreaction may therefore be used to make the water-in-oil emulsion, and serve to further enhance the stability of the resulting emulsion.

Tradenames in References

Tradename / Description	Supplier
Aerosil™ (Series)	Evonik Goldschmidt GmbH
Fumed silica (Bragg, 2000)	
CAB-O-SIL™ (Series)	Cabot. Corp.
Fumed silica (Bragg, 2000)	
Tegopren™ (Series)	Evonik Goldschmidt GmbH
Siloxane emulsifier (Jones and Wentzler, 2008)	
Tomadol®	Tomah Products, Inc.
Fatty amines (Jones and Wentzler, 2008)	

REFERENCES

Bragg, J.R., 2000. Oil recovery method using an emulsion. US Patent 6068054, assigned to Exxon Production Research Co., 30 May 2000.

Jones, T.A., Wentzler, T., 2008. Polymer hydration method using microemulsions. US Patent 7407915, assigned to Baker Hughes Inc., Houston, TX, 5 August 2008. <http://www.freepatentsonline.com/7407915.html>.

Kiel, O.M., 1973. Method of fracturing subterranean formations using oil-in-water emulsions. US Patent 3710865, assigned to Esso Production Research Co., 16 January 1973. <http://www.freepatentsonline.com/3710865.html>.

Kokal, S.L., 2006. Crude oil emulsions. In: Fanchi, J.R. (Ed.), Petroleum Engineering Handbook. vol. I. Society of Petroleum Engineers, Richardson, TX, pp. 533–570 (Chapter 12). <http://balikpapan.spe.org/images/balikpapan/articles/51/Crude%20Oil%20Emulsions%20-% 20PEH%20-Chapter%2012%20-%20compressed%20version.pdf>.

Kokal, S.L., Wingrove, M., 2000. Emulsion separation index: from laboratory to field case studies. In: Proceedings Volume. No. 63165-MS. SPE Annual Technical Conference and Exhibition, 1–4 October 2000, Society of Petroleum Engineers, Dallas, Texas.

Menon, V.B., Wasan, D.T., 1988. Characterization of oil-water interfaces containing finely divided solids with applications to the coalescence of water-in-oil emulsions: a review. Colloids Surf. 29 (1), 7–27. <http://www.sciencedirect.com/science/article/ B6W92-44CRG9F-3/2/082bfad854dc70aa811b9ca060b31a1e>.

Santanna, V.C., de Castro Dantas, T.N., Dantas Neto, A.A., 2012. The use of microemulsion systems in oil industry. In: Najjar, R. (Ed.), Microemulsions—An Introduction to Properties and Applications. InTech Europe, Rijeka, Croatia, pp. 161–174 (Chapter 8). <http://www.intechopen.com/books/microemulsions-an-introduction-to-properties-and-applications/the-use-of-microemulsion-systems-in-oil-industry>.

Sullivan, P.F., Tustin, G.J., Christanti, Y., Kubala, G., Drochon, B., Hughes, T.L., 2010. Aqueous two-phase emulsion gel systems for zone isolation. US Patent 7703527, assigned to Schlumberger Technology Corporation, Sugar Land, TX, 27 April 2010. <http://www.freepatentsonline.com/7703527.html>.

Taylor, R.S., Funkhouser, G.P., Dusterhoft, R.G., 2009. Gelled invert emulsion compositions comprising polyvalent metal salts of an organophosphonic acid ester or an organophosphinic acid and methods of use and manufacture. US Patent 7534745, assigned to Halliburton Energy Services, Inc., Duncan, OK, 19 May 2009. <http://www.freepatentsonline.com/7534745.html>.

Varadaraj, R., 2010. Oil-in-water-in-oil emulsion. US Patent 7652074, assigned to ExxonMobil Upstream Research Co., Houston, TX, 26 January 2010. <http://www.freepatentsonline.com/7652074.html>.

Varadaraj, R., Bragg, J.R., Dobson, M.K., Peiffer, D.G., Huang, J.S., Siano, D.B., Brons, C.H., Elspass, C.W., 2004. Solids-stabilized water-in-oil emulsion and method for using same. US Patent 6734144, assigned to ExxonMobil Upstream Research Co., Houston, TX, 11 May 2004. <http://www.freepatentsonline.com/6734144.html>.

Varadaraj, R., Bragg, J.R., Peiffer, D.G., Elspass, C.W., 2007. Stability enhanced water-in-oil emulsion and method for using same. US Patent 7186673, assigned to ExxonMobil Upstream Research Co., Houston, TX, 6 March 2007. <http://www.freepatentsonline.com/7186673.html>.

Yang, J., Jovancicevic, V., 2009. Microemulsion containing oil field chemicals useful for oil and gas field applications. US Patent 7615516, assigned to Baker Hughes Inc., Houston, TX, 10 November 2009. <http://www.freepatentsonline.com/7615516.html>.

Chapter 7

Demulsifiers

The class of demulsifiers may be required in certain systems to transform the viscous emulsion to a demulsified, low-viscosity state for promoting well cleanup (Crowell, 1984). However, in some systems, it may prove satisfactory to merely degrade the polymer. An emulsion without the polymer may have sufficient mobility to permit rapid well cleanup.

Emulsion fracturing fluids get their high viscosity from the dispersion of a major proportion of an oil-internal phase in a minor proportion of an external aqueous phase. The emulsion must be stabilized by a surfactant. To reduce the viscosity of the emulsion to permit easy removal of the fracturing fluid, it is necessary to break the emulsion into a water-in-oil invert or into its component phases.

The emulsion is usually broken by eliminating the stabilizing effect of the surfactant. This is normally accomplished either by adsorption of surfactant on the formation walls or by the addition of a demulsifying agent. Usually, only cationic surfactants are susceptible to adsorption on to formation materials because of their affinity for sand surfaces. If, on the other hand, a demulsifier is used, the demulsifier and surfactant must be carefully matched so that the emulsion begins to break only after fracturing is completed. Even with suitably matched demulsifiers and surfactants, it is not easy to accurately time the breaking of the emulsion and there is always the danger that demulsification will either take place prematurely or will be delayed for an unacceptably long period of time (Graham et al., 1976).

Although a good fracturing fluid, a polymer emulsion is especially difficult to remove from the formation because both the polymer and the emulsion significantly contribute to the high viscosity of the fluid. Consequently, mechanisms are needed to reduce the viscosity contribution of both. The polymer emulsion is usually made to convert to a low-viscosity fluid by a combined demulsification and polymer degradation system. The breaker system must permit completion of the fracturing treatment before acting and then must act quick enough to minimize recovery time. Related polymer problems with the polymer, such as degradation residues and ionic sensitivity, may also occur when a polymer emulsion is used.

Hydraulic Fracturing Chemicals and Fluids Technology. http://dx.doi.org/10.1016/B978-0-12-411491-3.00007-8
© 2013 Elsevier Inc. All rights reserved.

BASIC ACTION OF DEMULSIFIERS

Desired Properties

Demulsifiers for crude oil emulsions should meet the following properties:

- Rapid breakdown into water and oil with minimal amounts of residual water,
- Good shelf life, and
- Quick preparation.

Mechanisms of Demulsification

Stabilization of Water-Oil Emulsions

The stabilization of water-oil emulsions happens as a result of the interfacial layers, which mainly consist of colloids present in the crude oil-asphaltenes and resins. By adding demulsifiers, the emulsion breaks up. With water-soluble demulsifiers, the emulsion stabilizers originally in the system will be displaced from the interface. In addition, a change in wetting by the formation of inactive complexes may occur. Conversely, using oil-soluble demulsifiers, the mechanism, in addition to the displacement of crude colloids, is based on neutralizing the stabilization effect by additional emulsion breakers and the breakup resulting from interface eruptions (Kotsaridou-Nagel and Kragert, 1996).

Interfacial Tension Relaxation

The effectiveness of a crude oil demulsifier is correlated with the lowering of the shear viscosity and the dynamic tension gradient of the oil-water interface. The interfacial tension (IFT) relaxation occurs faster with an effective demulsifier (Tambe et al., 1995). Short relaxation times imply that IFT gradients at slow film thinning are suppressed. Electron spin resonance experiments with labeled demulsifiers indicate that the demulsifiers form reverse micellelike clusters in the bulk oil (Mukherjee and Kushnick, 1987). The slow unclustering of the demulsifier at the interface appears to be the rate-determining step in the tension relaxation process.

CHEMICALS

Demulsifiers in fracturing fluids are typically used in 0.1–0.5% v/v within a fracturing fluid (Crews, 2006). Examples of classes of demulsifier chemicals that are commonly used are shown in Table 7.1.

Further, certain chemicals are known to enhance the performance of demulsifiers. Various demulsifier enhancers include alcohols, aromatics, alkanolamines, carboxylic acids, amino carboxylic acids, bisulfites, hydroxides, sulfates, phosphates, polyols, and mixtures therefrom (Crews, 2006).

TABLE 7.1 Classes of Demulsifiers (Crews, 2006)

Class
Alkyl sulfates and sulfonates
Alkyl phosphonates
Alkyl quaternary amines and amine oxides
Oxyalkylated polyalkylene poly(amine)s
Fatty acid polyalkyl aromatic ammonium chloride
Poly(alkylene glycol)s and ethers
Poly(acrylate)s and acrylamides
Alkyl phenol resins
Oligoamine alkoxylates
Alkoxylated carboxylic acid esters
Ethoxylated alcohols
Organic and inorganic aluminum salts
Copolymers of acrylates-surfactants
Copolymers of acrylates-resins
Copolymers of acrylates-alkyl aromatic amines
Copolymers of carboxylics-polyols
Copolymers or terpolymers of alkoxylates with vinyl compounds
Condensates of monoamine or oligoamine alkoxylates
Dicarboxylic acids and alkylene oxide block copolymers

Alkyl ether organic acid esters or polyols may be included in treatment fluids used in hydrocarbon production to provide a demulsifying and defoaming action on foams and emulsions in the producing formation (Leshchyshyn et al., 2012; Giffin, 2012). The action of the compositions is time and temperature dependent and therefore their action can be controlled in situ. Special compositions have been described that contain demulsifier chemicals (Crews, 2007; Manz et al., 2005).

CHELATING AGENTS

Biodegradable and nontoxic chelating agent compositions can perform multiple beneficial functions in an aqueous fracturing fluid through the chelation of ions. Some of the multiple functions include:

- Demulsifier,
- Demulsifier enhancer,
- Scale inhibitor,
- Crosslink delay agent,

TABLE 7.2 Chelating Agents (Crews, 2006)

Compound
Sodium polyaspartate
Sodium iminodisuccinate
Disodium hydroxyethyleneiminodiacetate
Sodium gluconate
Sodium glucoheptonate
Sugar alcohols
Monosaccharides
Disaccharides

- Crosslinked gel stabilizer, and
- Enzyme breaker stabilizer.

Suitable chelating agents that are used in such compositions are shown in Table 7.2.

Tradenames in References

Tradename	Supplier
Description	
Aldacide® G	Halliburton Energy Services, Inc.
Biocide, glutaraldehyde (Giffin, 2012)	
Envirogem®	Air Products and Chemicals, Inc.
Nonionic surfactants (Giffin, 2012)	
Rhodoclean™	Rhodia Inc. Corp.
Nonionic surfactant (Giffin, 2012)	

REFERENCES

Crews, J.B., 2006. Biodegradable chelant compositions for fracturing fluid. US Patent 7078370. assigned to Baker Hughes Incorporated, Houston, TX, 18 July 2006. <http://www.freepatentsonline.com/7078370.html>.

Crews, J.B., 2007. Fracturing fluids for delayed flow back operations. US Patent 7256160. assigned to Baker Hughes Incorporated, Houston, TX, 14 August 2007. <http://www.freepatentsonline.com/7256160.html>.

Crowell, R.F., 1984. Formation fracturing method. US Patent 4442897. assigned to Standard Oil Company, Chicago, IL, 17 April 1984. <http://www.freepatentsonline.com/4442897.html>.

Giffin, W.J., 2012. Compositions and processes for fracturing subterranean formations. US Patent 8 293 687. assigned to Titan Global Oil Services Inc., Bloomfield Hills, MI, 23 October 2012. <http://www.freepatentsonline.com/8293687.html>.

Graham, J.W., Gruesbeck, C., Salathiel, W.M., 1976. Method of fracturing subterranean formations using oil-in-water emulsions. US Patent 3 977 472. assigned to Exxon Production Research Company, Houston, TX, 31 August 1976. <http://www.freepatentsonline.com/3977472.html>.

Kotsaridou-Nagel, M., Kragert, B., 1996. Demulsifying water-in-oil-emulsions through chemical addition (Spaltungsmechanismus von Wasser-in-Erdöl-Emulsionen bei Chemikalienzusatz). Erdöl Erdgas Kohle 112 (2), 72–75.

Leshchyshyn, T.T., Beaton, P.W., Coolen, T.M., 2012. Hydrocarbon-based fracturing fluid compositions, methods of preparation and methods of use. US Patent 8 211 834. assigned to Calfrac Well Services Ltd. Alberta, CA, 3 July 2012. <http://www.freepatentsonline.com/8211834.html>.

Manz, D.H., Mahmood, T., Khanam, H.A., 2005. Oil and gas well fracturing (frac) water treatment process. US Patent Application 20050098504. assigned to Davnor Water Treatment Technologies Ltd., Calgary, CA, 12 May 2005. <http://www.freepatentsonline.com/20050098504.html>.

Mukherjee, S., Kushnick, A.P., 1987. Effect of demulsifiers on interfacial properties governing crude oil demulsification. In: Proceedings Volume. Annual Aiche Meeting, New York, 15–20 November 1987.

Tambe, D., Paulis, J., Sharma, M.M., 1995. Factors controlling the stability of colloid-stabilized emulsions: Pt.4: evaluating the effectiveness of demulsifiers. J. Colloid Interface Sci. 171 (2), 463–469.

Chapter 8

Clay Stabilization

Problems caused by shales in petroleum activities are occurring frequently. At the beginning of the 1950s, many soil mechanics experts were interested in the swelling of clays. It is important to maintain the wellbore stability during drilling as well as during fracturing, especially in water-sensitive shale and clay formations.

The rocks within these types of formations absorb the fluid used in fracturing. This absorption causes the rock to swell and may lead to a wellbore collapse. The swelling of clays and the problems that may arise from these phenomena have been reviewed in the literature (Durand et al., 1995a,b; Zhou et al., 1995; Van Oort, 1997; Conway et al., 2011; Patel and Patel, 2012). Various additives for clay stabilization are shown in Table 8.1.

PROPERTIES OF CLAYS

Clay minerals are generally crystalline in nature. The structure of the clay crystals determines its properties. Typically, clays have a flaky, mica-type structure. Clay flakes are made up of a number of crystal platelets stacked face to face. Each platelet is called a unit layer, and the surfaces of the unit layer are called basal surfaces. A unit layer is composed of multiple sheets. One sheet type is called the octahedral sheet. It is composed of either aluminum or magnesium atoms octahedrally coordinated with the oxygen atoms of hydroxyl groups. Another sheet type is called the tetrahedral sheet. The tetrahedral sheet consists of silicon atoms tetrahedrally coordinated with oxygen atoms. Sheets within a unit layer link together by sharing oxygen atoms.

When this linking occurs between one octahedral and one tetrahedral sheet, one basal surface consists of exposed oxygen atoms while the other basal surface has exposed hydroxyl groups. It is also quite common for two tetrahedral sheets to bond with one octahedral sheet by sharing oxygen atoms. The resulting structure, known as the Hoffmann structure, has an octahedral sheet that is sandwiched between the two tetrahedral sheets (Hoffmann and Lipscomb, 1962). As a result, both basal surfaces in a Hoffmann structure are composed of exposed oxygen atoms.

Hydraulic Fracturing Chemicals and Fluids Technology. http://dx.doi.org/10.1016/B978-0-12-411491-3.00008-X
© 2013 Elsevier Inc. All rights reserved.

TABLE 8.1 Types of Clay Stabilizers

Additiv Type	References
Polymer lattices	Stowe et al. (2002)
Copolymers of anionic and cationic monomers	
Hydroxyaldehydes or hydroxyketones	Westerkamp et al. (1991)
Polyols and alkaline salt	Hale and van (1997)
Quaternary ammonium compounds	
In situ crosslinking of epoxide resins	Coveney et al. (1999a,b)
Quaternary ammonium carboxylates[BD, LT]	Himes (1992)
Quaternized trihydroxyalkyl amine[LT]	Patel and McLaurine (1993)
Poly(vinyl alcohol), potassium silicate, and potassium carbonate	Alford (1991)
Copolymer of styrene and substituted maleic anhydride (MA)	Smith and Balson (2000)
Potassium salt of carboxymethyl cellulose	Palumbo et al. (1989)
Water-soluble polymers with sulfosuccinate derivative-based surfactants, zwitterionic surfactants[BD, LT]	Alonso-Debolt and Jarrett (1995,1994)

BD: Biodegradable, LT: Low toxicity, SF: Well stimulation fluid.

The unit layers stack together face to face and are held in place by weak attractive forces. The distance between corresponding planes in adjacent unit layers is called the c-spacing. A clay crystal structure with a unit layer consisting of three sheets typically has a c-spacing of about 9.5×10^{-7} mm.

In clay mineral crystals, atoms having different valences commonly will be positioned within the sheets of the structure to create a negative potential at the crystal surface. In that case, a cation is adsorbed on the surface. These adsorbed cations are called exchangeable cations, because they may chemically trade places with other cations when the clay crystal is suspended in water. In addition, ions may also be adsorbed on the clay crystal edges and exchange with other ions in the water.

Swelling of Clays

In clay mineral crystals, atoms with different valences will be commonly positioned within the sheets of the structure to create a negative potential at the crystal surface. In that case, a cation is adsorbed on the surface. These adsorbed cations are called exchangeable cations because they may chemically trade places with other cations when the clay crystal is suspended in water. In addition, ions may also be adsorbed on the clay crystal edges and exchange with other ions in the water (Patel et al., 2007).

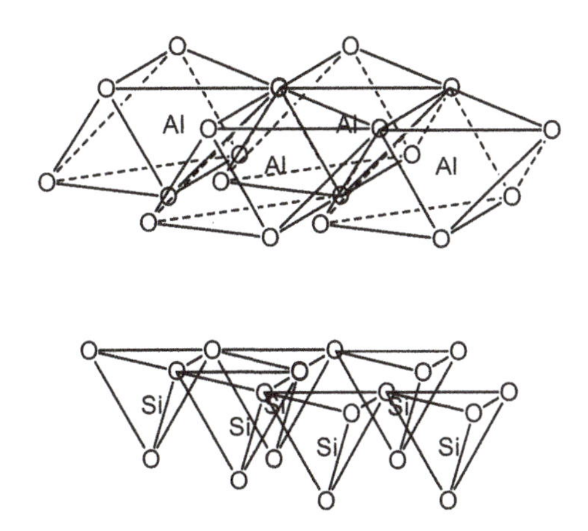

FIGURE 8.1 Octahedral and tetrahedral layers in clays (Murray, 2007, p. 9).

The type of substitutions occurring within the clay crystal structure and the exchangeable cations adsorbed on the crystal surface greatly affect the clay swelling. Clay swelling is a phenomenon in which water molecules surround a clay crystal structure and position themselves to increase the structure's c-spacing thus resulting in an increase in volume. Two types of swelling may occur (Patel et al., 2007):

Surface hydration is one type of swelling in which water molecules are adsorbed on crystal surfaces. Hydrogen bonding holds a layer of water molecules to the oxygen atoms exposed on the crystal surfaces. Subsequent layers of water molecules align to form a quasi-crystalline structure between unit layers which results in an increased c-spacing. All types of clays swell in this manner.

Osmotic swelling is a second type of swelling. Where the concentration of cations between unit layers in a clay mineral is higher than the cation concentration in the surrounding water, water is osmotically drawn between the unit layers and the c-spacing is increased. Osmotic swelling results in larger overall volume increases than surface hydration. However, only a few clays, like sodium montmorillonite, swell in this manner (Patel et al., 2007).

Clays are naturally occurring layered minerals which are formed by weathering and decomposition of igneous rocks. Details of clay mineralogy can be found in monographs (Grim, 1968; Murray, 2007). Each layer is comprised of fused sheets of octahedra of Al^{3+},Mg^{2+}, or Fe^{3+} oxides and sheets of tetrahedra of Si^{4+} oxides (Auerbach, 2007). If a clay mineral is constituted from one tetrahedral and one octahedral sheet, it is addressed as a 1:1 clay. If a clay contains two tetrahedral sheets sandwiching one central octahedral sheet, it is called a 2:1 clay.

Octahedral and tetrahedral layers are illustrated in Figure 8.1. The metal atoms in the clay lattice can be substituted appropriately which results in an overall negative charge on individual clay layers.

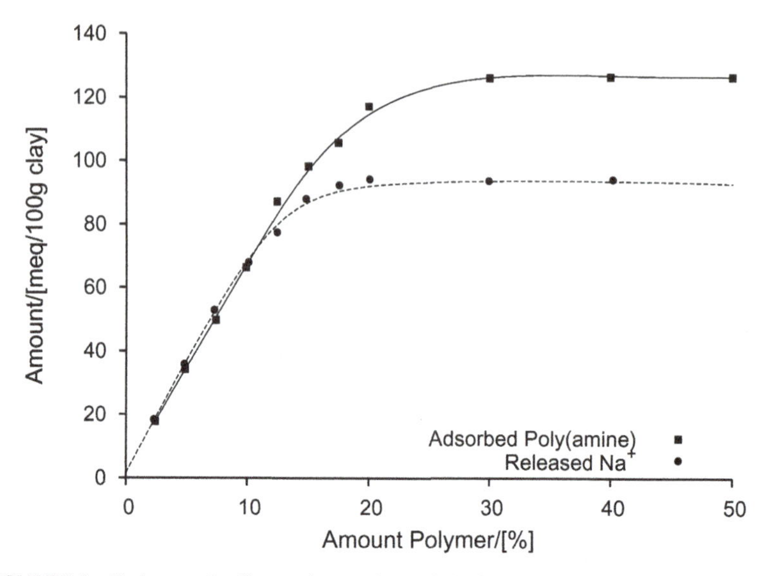

FIGURE 8.2 Exchange of sodium cations against poly(amine) cations (Blachier et al., 2009).

This charge is compensated for by cations located in the interlayer region. The latter cations can be freely exchanged. The cation exchange capacity of clay minerals depends on the crystal size, pH, and the type of the cation. These cations may not only be small-sized ions, but may be even polycations (Blachier et al., 2009).

Studies on the adsorption of a polycationic quaternary amine polymer on clays have been presented. In charge scale, it can be observed that both the adsorption curve of the quaternary amine polymer and that corresponding to the released sodium are superimposed, as shown in Figure 8.2. The replacement of the counterions by the amine polymer almost follows a 1:1 relationship for low polymer concentrations. Further, the silicate surfaces of the tetrahedral sheets of clay minerals are comparatively hydrophobic. This property may allow the intercalation of neutral organic compounds including polymers.

Smectite clays are of the type 2:1 and are frequently occurring (Anderson et al., 2010). Sodium-saturated smectite swells macroscopically. This swelling causes the instability of shales in oil well operations. In the worst case, the wellbore may collapse due to clay swelling.

Exchangeable cations found in clay minerals are reported to have a significant impact on the amount of swelling that takes place. The exchangeable cations compete with water molecules for the available reactive sites in the clay structure. Generally, cations with high valences are more strongly adsorbed than those with low valences. Thus, clays with low valence exchangeable cations will swell more than clays whose exchangeable cations have high valences.

Minimizing the clay swelling is an important field of research. In order to reduce the extent of clay swelling effectively, the mechanism of swelling needs to be understood. Based on this knowledge, efficient swelling inhibitors may be developed.

Suitable clay swelling inhibitors must both significantly reduce the hydration of the clay and must meet the increasingly stringent environmental guidelines.

Swelling takes place in a discrete fashion, namely in the stepwise formation of integer-layer hydrates. The transitions of the distances of the layers are thermodynamically analogous to phase transitions. Electro-osmotic swelling can occur only in clay minerals that contain exchangeable cations in the interlayer region. This type of swelling may yield significantly larger expansion than crystalline swelling.

Sodium-saturated smectites have a strong tendency toward electro-osmotic swelling. In contrast, potassium-saturated smectites do not swell in such a manner. Thus allowing an appropriate ion exchange reaction can be helpful in clay stabilization (Anderson et al., 2010).

The water desorption isotherms of montmorillonite intercalated with exchangeable cations of the alkali metal group showed that for larger cations, less water is adsorbed (Mooney et al., 1952). In addition, there is a relationship with the tendency to swell and the energy of hydration of the cation (Norrish, 1954). Gumbo clay is notorious for its swelling (Klein and Godinich, 2006).

Montmorillonite

Montmorillonite clays, for example Bentonite and Kaolinite clays, are also suitable for preparing a solids-stabilized oil-in-water emulsion. Bentonite clay can be easily exfoliated (Bragg and Varadaraj, 2006). As mined, bentonite clays naturally consist of aggregates of particles that can be dispersed in water and broken up by shearing into units having average particle sizes of 2 μm or less. However, each of these particles is a laminated unit containing approximately 100 layers of fundamental silicate layers of 1 nm thickness bonded together by inclusions of atoms such as calcium in the layers.

By exchanging the atoms such as calcium by sodium or lithium, which are larger and have strong attractions for water molecules in fresh water, and then exposing the bentonite to fresh water, the bentonite can be broken into individual 1 nm thick layers, called fundamental particles. The result of this delamination process is a gel consisting of a finely divided bentonite clay (Bragg and Varadaraj, 2006).

Guidelines

The literature offers several papers that may serve as guidelines for issues such as selecting a proper clay stabilizing system or completing wellbore stability analysis of practical well designs (Chen et al., 1996; Crowe (1990, 1991a,b); Evans and Ali, 1997; Scheuerman and Bergersen, 1989).

MECHANISMS CAUSING INSTABILITY

Shale stability is an important problem faced during various well bore operations. Stability problems are attributed most often to the swelling of shales. It has been shown that several mechanisms can be involved (Gazaniol et al. (1994, 1995)). These can be pore pressure diffusion, plasticity, anisotropy, capillary effects, osmosis, and physicochemical alterations. Most important, three processes contributing to the instability of shales have to be considered (Bailey et al., 1994):

1. Movement of fluid between the wellbore and shale (limited to flow from the wellbore into the shale).
2. Changes in stress and strain that occur during shale-filtrate interaction.
3. Softening and erosion caused by invasion of mud filtrate and consequent chemical changes in the shale.

The major reason for these effects is due to the chemical nature, namely the hydration of clays. Borehole instabilities were observed even with the most inhibitive fluids, i.e., oil-based drilling muds. This demonstrates that the mechanical aspect is also important. In fact, the coupling of both chemical and mechanical mechanisms has to be considered. For this reason, it is still difficult to predict the behavior of rock at medium-to-great depth under certain loading conditions.

The stability of shales is governed by a complicated relationship between transport processes in shales (e.g., hydraulic flow, osmosis, diffusion of ions, pressure) and chemical changes (e.g., ion exchange, alteration of water content, swelling pressure).

Clays or shales have the ability to absorb water, thus causing the instability of wells either because of the swelling of some mineral species or because the supporting pressure is suppressed by modification of the pore pressure. The response of a shale to a water-based fluid depends on its initial water activity and on the composition of the fluid.

The behavior of shales can be classified into either deformation mechanisms or transport mechanisms (Tshibangu et al., 1996). Optimization of mud salinity, density, and filter cake properties is an important issue.

Kinetics of Swelling of Clays

Basic studies on the kinetics of swelling have been performed (Suratman, 1985). Pure clays (montmorillonite, illite, and kaolinite) with polymeric inhibitors were investigated, and phenomenologic kinetic laws were established.

Hydrational Stress

Stresses caused by chemical forces, such as hydration stress, can have a considerable influence on the stability of a wellbore (Chen et al., 1995). When the total pressure and the chemical potential of water increase, water is absorbed into the clay platelets.

This results either in the platelets moving farther apart (swelling) if they are free to move or in generation of hydrational stress if swelling is constrained (Tan et al., 1997). Hydrational stress results in an increase in pore pressure and a subsequent reduction in effective mud support, which leads to a less stable wellbore condition.

Borehole Stability Model

A borehole stability model has been developed that takes into account both the mechanical and the chemical aspects of interactions between drilling fluid and shale (Mody and Hale, 1993). Chemically induced stress alteration based on the thermodynamics of differences in water molar free energies of the drilling fluid and shale is combined with mechanically induced stress. Based on this model, it should be possible to obtain the optimal mud weight and salt concentration for drilling fluids.

Further stability models based on surface area, equilibrium water-content pressure relationships, and electric double-layer theory can successfully characterize borehole stability problems (Wilcox, 1990). The application of surface area, swelling pressure, and water requirements of solids can be integrated into swelling models and mud process control approaches to improve the design of water-based drilling mud in active or older shales.

Shale Inhibition with Water-Based Muds

One potential mechanism by which polymers may stabilize shales is by reducing the rate of water invasion into the shale. The control of water invasion is not the only mechanism involved in shale stabilization (Ballard et al., 1993). There is also an effect of the polymer additive. Osmotic phenomena are responsible for water transport rates through shales.

Inhibiting Reactive Argillaceous Formations

Argillaceous formations are very reactive in the presence of water. Such formations can be stabilized by bringing them in contact with a polymer solution with hydrophilic and hydrophobic links (Audibert et al. (1993 a,b)). The hydrophilic portion consists of poly(oxyethylene), with hydrophobic end groups based on isocyanates. The polymer is capable of inhibiting the swelling or dispersion of the argillaceous rock resulting from its adsorptive and hydrophobic capacities.

Formation Damage by Fluids

Formation damage due to invasion by drilling fluids is a well-known problem in drilling. Invasion of drilling fluids into the formation is caused by the differential pressure of the hydrostatic column which is generally greater than the formation pressure, especially in low pressure or depleted zones (Whitfill et al., 2005).

Invasion is also caused by openings in the rock and the ability of fluids to move through the rock. When drilling depleted sands under overbalanced conditions, the mud will penetrate progressively into the formation unless there is an effective flow barrier present at the wellbore wall.

SWELLING INHIBITORS

Inhibitors of swelling act in a chemical manner rather than in a mechanical manner. They change the ionic strength and the transport behavior of the fluids into the clays. Both the cations and the anions are important for the efficiency of the inhibition of swelling of clays (Doleschall et al., 1987).

Salts

Swelling can be inhibited by the addition of KCl. Relatively high levels are required. Other swelling inhibitors are both uncharged polymers and poly(electrolyte)s (Anderson et al., 2010).

Quaternary Ammonium Salts

Choline salts are effective antiswelling drilling fluid additives for underbalanced drilling operations (Kippie and Gatlin, 2009).

Choline is addressed as a quaternary ammonium salt containing the *N,N,N*-trimethylethanolammonium cation. An example of choline halide counterion salts is choline chloride.

Preparation 8.1. Triethanolamine methyl chloride can be prepared by adding to triethanol amine in aqueous solution methyl chloride in excess and heating for several hours. Upon completion of the reaction, the excess of methyl chloride is evaporated.

Choline formate is prepared from an aqueous solution of choline hydroxide by the reaction with formic acid simply by stirring. ■

Argillaceous formations contain clay particles. If a water-based drilling fluid is used in such formations, ion exchange, hydration, etc., will take place. These reactions cause swelling, crumbling, or dispersion of the clay particles. Ultimately, washout and even complete collapse of the borehole may occur (Eoff et al., 2006).

Certain additives may prevent these unfavorable reactions. These additives are essentially quaternized polymers. Such polymers have been shown in laboratory testing to vastly reduce shale erosion. Quaternized polymers can be synthesized by (Eoff et al., 2006):

1. Quaternization of an AA-based amine derivative with an alkyl halide, and subsequent polymerization, or
2. First polymerization and afterwards quaternization of the polymeric moieties.

Preparation 8.2. A quaternized monomer can be prepared by mixing dimethyl amino ethyl methacrylate with hexadecyl bromide. The mixture is heated to 43 °C and stirred for 24 h. Then, the mixture is poured into petroleum ether, whereby the quaternized monomer precipitates (Eoff et al., 2006). The reaction is shown in Figure 8.3. ■

A copolymer can be prepared using the quaternized monomer, described above and the dimethyl amino ethyl methacrylate as such. The aqueous solution is neutralized with sulfuric acid and radically polymerized with 2,2′-azobis(2-amidinopropane)dihydrochloride, c.f. Figure 8.4. This initiator is water soluble. The polymerization is carried out at 43 °C for 18 h (Eoff et al., 2006).

The quaternization of a polymer from dimethyl amino ethyl methacrylate has been described. To an aqueous solution of a homopolymer from dimethyl amino ethyl methacrylate, sodium hydrochloride is added to adjust the pH to 8.9. Then again some water is added and hexadecyl bromide as alkylation agent, further benzylcetyldimethyl ammonium bromide as emulsifier. This mixture is then heated, with stirring, to 60 °C for 24 h (Eoff et al., 2006).

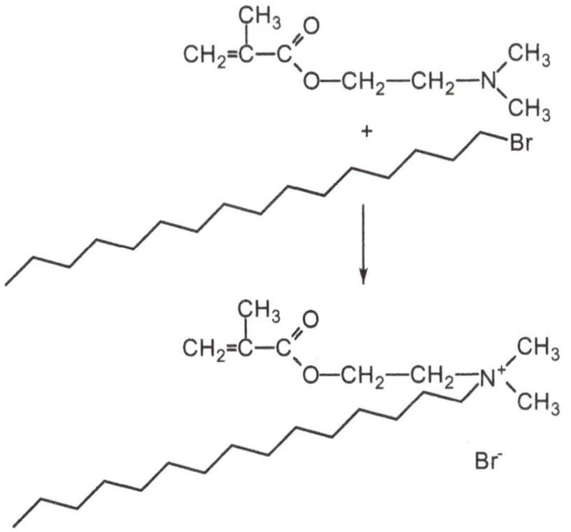

$CH_2{=}C(CH_3){-}C({=}O){-}O{-}CH_2{-}CH_2{-}N(CH_3)_2$ + $C_{16}H_{33}Br$ → $CH_2{=}C(CH_3){-}C({=}O){-}O{-}CH_2{-}CH_2{-}N^+(CH_3)_2(C_{16}H_{33})$ Br^-

FIGURE 8.3 Quaternization reaction of dimethyl amino ethyl methacrylate with hexadecyl bromide.

2 HCl

2,2′-Azobis (2-amidinopropane) dihydrochloride

FIGURE 8.4 Water-soluble radical initiator.

Potassium formate

Clay is stabilized in drilling and treatment operations by the addition of potassium formate to the drilling fluid. Further, a cationic formation control additive is added. Potassium formate can be generated in situ from potassium hydroxide and formic acid. The cationic additive is basically, a polymer containing quaternized amine units, e.g., polymers of dimethyl diallyl ammonium chloride or acrylamide (AAm) (Smith, 2009).

In the clay pack flow test, where the higher volumes at a given time indicate better clay stability, the addition of a small amount of potassium formate increases the volume throughput for a given polymer concentration. For example, 0.1% poly(dimethyl diallyl ammonium chloride) added to the formulation had a volume at 10 min of 112 ml.

The same polymer, when combined with potassium formate and treated at 0.05% of the polymer, i.e., half the original polymer concentration, had a volume of 146 ml, indicating better clay stability and a possible synergistic effect from the addition of the potassium formate (Smith, 2009).

Saccharide Derivatives

A fluid additive that acts as a clay stabilizer is the reaction product of methyl glucoside and alkylene oxides such as ethylene oxide (EO), propylene oxide (PO), or 1,2-butylene oxide. Such an additive is soluble in water at ambient conditions, but becomes insoluble at elevated temperatures (Clapper and Watson, 1996). Because of their insolubility at elevated temperatures, these compounds concentrate at important surfaces such as the drill bit cutting surface, the borehole surface, and the surfaces of the drilled cuttings.

Sulfonated Asphalt

Asphalt is a solid, black-brown to black bitumen fraction, which softens when heated and re-hardens upon cooling. Asphalt is not water soluble and difficult to disperse or emulsify in water.

Sulfonated asphalt can be obtained by reacting asphalt with sulfuric acid and sulfur trioxide. By neutralization with alkali hydroxides, such as NaOH or NH_3, sulfonate salts are formed. Only a limited portion of the sulfonated product can be extracted with hot water. However, the fraction thus obtained, which is water soluble, is crucial for the quality.

Sulfonated asphalt is commonly utilized for water-based and oil-based drilling fluids (Huber et al., 2009). The mechanism of action of sulfonated asphalt as a clay inhibitor is explained that the electronegative sulfonated macromolecules attach to the electropositive ends of the clay platelets. Thereby, a neutralization barrier is created, which suppresses the absorption of water into the clay.

In addition, because the sulfonated asphalt is partially lipophilic, and therefore water repellent, the water influx into the clay is restricted by purely physical principles. As mentioned already, the solubility in water of the sulfonated asphalt is crucial for proper application. By the introduction of a water-soluble and an anionic polymer component, the proportion of water-insoluble asphalt can be markedly reduced.

In other words, the proportion of the water-soluble fraction is increased by introducing the polymer component. Especially suitable are lignosulfonates as well as sulfonated phenol, ketone, naphthalene, acetone, and amino plasticizing resins (Huber et al., 2009).

Grafted Copolymers

The clay stabilization of copolymers of styrene and MA grafted with poly(ethylene glycol) (PEG) has been investigated (Smith and Balson, 2004).

The amounts of shale recovery from bottle rolling tests have been used to measure the shale inhibition properties. The tests were done using Oxford Clay cuttings, a water-sensitive shale, sieved to 2–4 mm. The swelling is performed in 7.6% aqueous KCl.

The grafted copolymer used is an alternating copolymer of styrene and MA. The polymer is grafted with PEG with different molecular weight. The amount of shale recovery with various PEG types is shown in Table 8.2.

It seems that there is an optimum, with respect to the molecular weight of the grafted PEG. Further, the results in the lower part of Table 8.2 indicate that increasing the amount of styrene in the backbone increases also the amount of shale recovered.

Poly(oxyalkylene amine)s

One method to reduce the swelling of a clay is to add salts to the fluids. Salts generally reduce the swelling of clays. However, salts flocculate the clays resulting in both high fluid losses and an almost complete loss of thixotropy. Further, increasing the salinity often decreases the functional characteristics of drilling fluid additives (Patel et al., 2007).

Another method for controlling clay swelling is to use organic shale inhibitor molecules in drilling fluids. It is believed that the organic shale inhibitor molecules are adsorbed on the surfaces of clays with the added organic shale

TABLE 8.2 Amount of Shale Recovery (Smith and Balson, 2004)

Sample	KCl (%)	Shale recovery (%)
KCl only	7.6	25
PEG	7.6	38
SMAC MPEG 200	7.6	54
SMAC MPEG 300	7.6	87
SMAC MPEG 400	7.6	85
SMAC MPEG 500	7.6	72
SMAC MPEG 600	7.6	69
SMAC MPEG 750	7.6	70
SMAC MPEG 1100	7.6	66
SMAC MPEG 1500	7.6	49
KCl only	12.9	27
PEG	12.9	53
SMAC MPEG 500	12.9	85
SMAC 2:1 MPEG 500	12.9	95

SMAC: Styrene and MA copolymer.

SMAC 2:1: Styrene and MA copolymer, 2 styrene units for every MA.

MPEG: Poly(ethylene glycol) monomethyl ethers, the number refers to the molecular weight.

inhibitor competing with water molecules for clay reactive sites and thus serve to reduce clay swelling.

Poly(oxyalkylene amine)s are a general class of compounds that contain primary amino groups attached to a poly(ether) backbone. They are also addressed as poly(ether amine)s. They are available in a variety of molecular weights, ranging up to 5 kDa.

Poly(oxyalkylenediamine)s have been proposed as shale inhibitors. These are synthesized from the ring opening polymerization of oxirane compounds in the presence of amino compounds. Such compounds have been synthesized by reacting Jeffamine with two equivalents of EO. Alternatively, PO is reacted with an oxyalkyldiamine (Patel et al., 2007). The poly(ether) backbone is based either on EO, or PO, or a mixture of these oxirane compounds (Patel et al., 2007).

A typical poly(ether amine) is shown in Figure 8.5. Such products belong to the Jeffamine® product family. A related shale hydration inhibition agent is based on an *N*-alkylated 2,2′-diaminoethylether.

Anionic Polymers

Anionic polymers may be active as the long chain with negative ions attaches to the positive sites on the clay particles or to the hydrated clay surface through

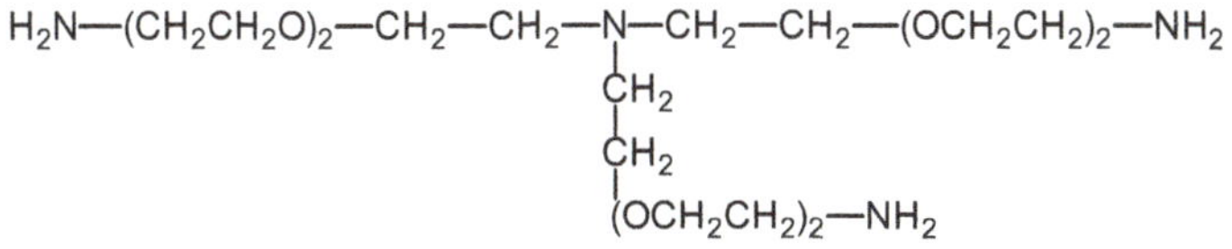

$$H_2N{-}(CH_2CH_2O)_2{-}CH_2{-}CH_2{-}\underset{\displaystyle \begin{array}{c} | \\ CH_2 \\ | \\ CH_2 \\ | \\ (OCH_2CH_2)_2{-}NH_2 \end{array}}{N}{-}CH_2{-}CH_2{-}(OCH_2CH_2)_2{-}NH_2$$

FIGURE 8.5 Poly(ether amine) (Klein and Godinich, 2006).

hydrogen bonding (Halliday and Thielen, 1987). Surface hydration is reduced as the polymer coats the surface of the clay.

The protective coating also seals, or restricts the surface fractures or pores, thereby reducing or preventing the capillary movement of filtrate into the shale. This stabilizing process is supplemented by PAC. Potassium chloride enhances the rate of polymer absorption onto the clay.

Amine Salts of Maleic Imide

Compositions containing amine salts of imides of MA polymers are useful for clay stabilization. These types of salts are formed, for example by the reaction of MA with a diamine such as dimethyl aminopropylamine, in ethylene glycol (EG) solution (Poelker et al., 2009). The primary nitrogen dimethyl aminopropylamine forms the imide bond.

In addition, it may add to the double bond of MA. Further, the EG may add to the double bond, but also may condense with the anhydride itself. On repetition of these reactions, oligomeric compounds may be formed. The elementary reactions are shown in Figure 8.6.

Finally, the product is neutralized with acetic acid or methanesulfonic acid to a pH of 4. The performance was tested in Bandera sandstone. The material neutralized with methanesulfonic acid is somewhat less than that neutralized with acetic acid. The compositions are particularly suitable for water-based hydraulic fracturing fluids.

Guanidyl Copolymer

The clay swelling and the migration of fines can be reduced by the addition of a guanidyl copolymer (Murphy et al., 2012). The guanidyl copolymer is the condensation product of an amino base, formaldehyde, an alkylenepolyamine, and the ammonium salt of an inorganic or organic acid. The basic procedure for the preparation has been reported (Waldmann, 1997 and is summarized in Preparation 8.3).

Preparation 8.3. Diethylenetriamine is heated to 55–60 °C with stirring, followed by the addition of ammonium chloride. Then the mixture is heated to 95–100 °C. Then diethylene glycol and also dicyandiamide is added. The

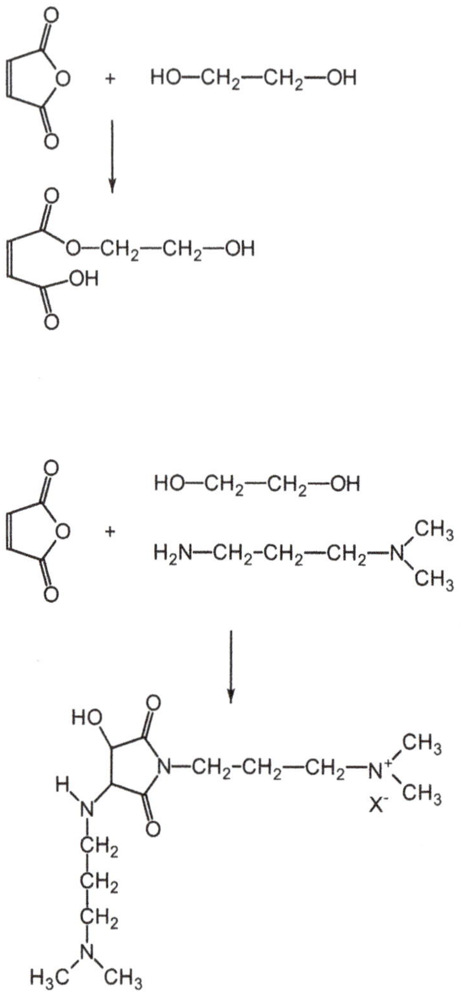

HO—CH$_2$—CH$_2$—OH

O—CH$_2$—CH$_2$—OH

OH

HO—CH$_2$—CH$_2$—OH

H$_2$N—CH$_2$-CH$_2$—CH$_2$—N(CH$_3$)$_2$

FIGURE 8.6 Start of condensation with ethylene glycol (top) and formation of amine salts of imides (bottom) (Poelker et al., 2009).

reactor temperature is gradually increased up to 195 °C and afterwards cooled by a specific protocol (Waldmann, 1997). ■

Variations of the monomers are shown in Table 8.3. The guanidyl copolymer has preferably a weight average molecular weight of 1–20 kDa.

The combination of the guanidyl copolymer and a cationic friction reducer allows for significant reductions in the amount of friction reducer employed.

TABLE 8.3 Monomers for Guanidyl Copolymers (Murphy et al., 2012)

Amine Reactant	Carbonyl Reactant	Amine Reactant
Dicyanodiamine	Formaldehyde	Aminoethylpiperizine
Guanamine	Paraformaldehyde	Ethylenediamine
Guanidine	Urea	Diethylenetriamine
Melamine	Thiourea	Triethylenetetramine
Cyanamine	Glyoxal	Propylenediamine
Guanylurea	Acetaldehyde	Tributylenetetramine
Guanylthiourea	Propionaldehyde	Tetrabutylenepentamine
Alkyl guanidine	Butyraldehyde	Dipentylenetriamine
Aryl guanidine	Glutaraldehyde	Tripentylenetetramine
	Acetone	Pentylenediamine

For example, the guanidyl copolymer allows for a reduction of 30–70%, in the total dosage of the friction reducer added to the fracturing fluid (Murphy et al., 2012). This can lead to significant cost savings over traditional compositions.

TABLE 8.4 Special Clay Stabilizers

Compound	References
Ammonium chloride	
Potassium chloride[a]	Yeager and Bailey (1988)
Dimethyl Diallyl ammonium salt[b]	Thomas and Smith (1993)
N-Alkyl pyridinium halides	
N,N,N-Trialkylphenylammonium halides	
N,N-Dialkylmorpholinium halides[c]	Himes (1992) and Himes and Vinson (1989)
Reaction product of a homopolymer of MA and an alkyl diamine[d]	Schield et al. (1991)
Tetramethylammonium chloride and methyl chloride	Aften and Gabel (1992)
Quaternary salt of ethylene-ammonia condensation polymer[d]	
Quaternary ammonium compounds[e]	Hall and Szememyei (1992)

[a]*Added to a gel concentrate with a diesel base.*
[b]*Minimum 0.05% to prevent swelling of clays.*
[c]*Alkyl equals methyl, ethyl, propyl, and butyl.*
[d]*Synergistically retards water absorption by the clay formation.*
[e]*Hydroxyl-substituted alkyl radials.*

Special Clay Stabilizers

Advances in clay-bearing formation treatment have led to the development of numerous clay stabilizing treatments and additives. Most additives used are high-molecular-weight cationic organic polymers. However, it has been shown that these stabilizers are less effective in low-permeability formations (Himes et al., 1989).

The use of salts, such as potassium chloride and sodium chloride, as temporary clay stabilizers during oil well drilling, completion, and servicing, has been practiced for many years. Because of the bulk and potential environmental hazards associated with the salts, many operators have looked for alternatives.

Recent research has shown a relationship between physical properties of various cations (e.g., K^+, Na^+) and their efficiency as temporary clay stabilizers. These properties were used to synthesize an organic cation (Table 8.4) with a higher efficiency as a clay stabilizer.

The additives provide additional benefits when used in conjunction with acidizing and fracturing treatments, since a lower salt concentration can be used (Himes et al., 1990; Himes and Vinson, 1991). The liquid product has been proven to be much easier to handle and transport. It is environmentally compatible and biodegradable in its diluted form.

Tradenames in References

Tradename	Supplier
Description	
Aerosil®	Degussa AG
Fumed Silica (Bragg and Varadaraj, 2006)	
Carbolite™	Carbo Corp.
Sized ceramic proppant (Kippie and Gatlin, 2009)	
Dacron®	DuPont
Poly(ethylene terephthalate) (Kippie and Gatlin, 2009)	
Jeffamine® D-230	Huntsman
Poly(oxypropylene) diamine (Klein and Godinich, 2006)	
Jeffamine® EDR-148	Huntsman
Triethyleneglycol diamine (Klein and Godinich, 2006)	
Jeffamine® HK-511	Huntsman
Poly(oxyalkylene) amine (Klein and Godinich, 2006)	
Shale Guard™ NCL100	Weatherford Int.
Shale anti-swelling agent (Kippie and Gatlin, 2009)	

REFERENCES

Aften, C.W., Gabel, R.K., 1992. Clay stabilizing method for oil and gas well treatment. US Patent 5 099 923, assigned to Nalco Chemical Co., 31 March 1992.

Alford, S.E., 1991. North Sea field application of an environmentally responsible water-base shale stabilizing system. In: Proceedings Volume. SPE/IADC Drilling Conference, Amsterdam, the Netherland, 11–14 March 1991, pp. 341–355.

Alonso-Debolt, M.A., Jarrett, M.A., 1994. New polymer/surfactant systems for stabilizing troublesome gumbo shale. In: Proceedings Volume. SPE International Petroleum Conference of Mexico, Veracruz, Mexico, 10–13 October 1994, pp. 699–708.

Alonso-Debolt, M., Jarrett, M., 1995. Synergistic effects of sulfosuccinate/polymer system for clay stabilization. In: Proceedings Volume. Vol. PD-65. Asme Energy-Sources Technological Conference Drilling Technological Symposium, Houston, 29 January –1 February 1995, pp. 311–315.

Anderson, R.L., Ratcliffe, I., Greenwell, H.C., Williams, P.A., Cliffe, S., Coveney, P.V., 2010. Clay swelling—A challenge in the oilfield. Earth-Sci. Rev. 98 (3–4), 201–216. <http://www.sciencedirect.com/science/article/B6V62-4XRCRVK-1/2/2ff8baefa17f8a009368b42d7f32f3ad>.

Audibert, A., Lecourtier, J., Bailey, L., Maitland, G., 1993a. Method for inhibiting reactive argillaceous formations and use thereof in a drilling fluid. WO Patent 9 315 164, assigned to Schlumberger Technol. Corp., Schlumberger Serv. Petrol, and Inst. Francais Du Petrole, 5 August 1993.

Audibert, A., Lecourtier, J., Bailey, L., Maitland, G., 1993b. Process for inhibiting reactive argillaceous formations and application to a drilling fluid (procede d'inhibition de formations argileuses reactives et application a un fluide de forage). FR Patent 2 686 892, assigned to Inst. Francais Du Petrole and Schlumberger Cambridge Re, 6 August 1993.

Audibert, A., Lecourtier, J., Bailey, L.C., Maitland, G., 1994. Use of polymers having hydrophilic and hydrophobic segments for inhibiting the swelling of reactive argillaceous formations (l'utilisation d'un polymere en solution aqueuse pour l'inhibition de gonflement des formations argileuses reactives). EP Patent 578 806, assigned to Schlumberger Serv. Petrol and Inst. Francais Du Petrole, 19 January 1994.

Audibert, A., Lecourtier, J., Bailey, L., Maitland, G., 1997. Method for inhibiting reactive argillaceous formations and use thereof in a drilling fluid. US Patent 5 677 266, assigned to Inst. Francais Du Petrole, 14 October 1997.

Auerbach, S.M. (Ed.), 2007. Handbook of layered materials, reprint from 2004 edition. CRC Press, Boca Raton.

Bailey, L., Reid, P.I., Sherwood, J.D., 1994. Mechanisms and solutions for chemical inhibition of shale swelling and failure. In: Proceedings Volume. Recent Advances in Oilfield Chemistry, 5th Royal Society of Chemistry. International Symposium, Ambleside, England, 13–15 April 1994, pp. 13–27.

Ballard, T., Beare, S., Lawless, T., 1993. Mechanisms of shale inhibition with water based muds. In: Proceedings Volume. IBC Technical Services Ltd., Preventing Oil Discharge from Drilling Operation—The Options Conference, Aberdeen, Scotland, 23–24 June 1993.

Blachier, C., Michot, L., Bihannic, I., Barrès, O., Jacquet, A., Mosquet, M., 2009. Adsorption of polyamine on clay minerals. J. Colloid Interface Sci. 336 (2), 599–606. <http://www.sciencedirect.com/science/article/B6WHR-4W2NDRN-B/2/0cdcd6ee0039c2579d4807db98e548d5>.

Bragg, J.R., Varadaraj, R., 2006. Solids-stabilized oil-in-water emulsion and a method for preparing same. US Patent 7 121 339, assigned to ExxonMobil Upstream Research Company, Houston, TX, 17 October 2006. <http://www.freepatentsonline.com/7121339.html>.

Chen, M., Chen, Z., Huang, R., 1995. Hydration stress on wellbore stability. In: Proceedings Volume. 35th US Rock Mech Symposium, Reno, NV, 5–7 June 1995, pp. 885–888.

Chen, X., Tan, C.P., Haberfield, C.M., 1996. Wellbore stability analysis guidelines for practical well design. In: Proceedings Volume. SPE Asia Pacific Oil & Gas Conference, Adelaide, Australia, 28–31 October 1996, pp. 117–126.

Clapper, D.K., Watson, S.K., 1996. Shale stabilising drilling fluid employing saccharide derivatives. EP Patent 702 073, assigned to Baker Hughes Inc., 20 March 1996.

Conway, M., Venditto, J.J., Reilly, P., Smith, K., 2011. An examination of clay stabilization and flow stability in various north american gas shales. In: Proceedings of SPE Annual Technical Conference and Exhibition. SPE Annual Technical Conference and Exhibition, 30 October–2 November 2011, Denver, Colorado, USA, Society of Petroleum Engineers, Dallas, Texas, pp. 1–17. http://dx.doi.org/10.2118/147266-MS.

Coveney, P.V., Watkinson, M., Whiting, A., Boek, E.S., 1999a. Stabilising clayey formations. GB Patent 2 332 221, assigned to Sofitech NV, 16 June 1999.

Coveney, P.V., Watkinson, M., Whiting, A., Boek, E.S., 1999b. Stabilizing clayey formations. WO Patent 9 931 353, assigned to Sofitech NV, Dowell Schlumberger SA, and Schlumberger Canada Ltd., 24 June 1999.

Crowe, C.W., 1990. Laboratory study provides guidelines for selecting clay stabilizers. In: Proceedings Volume. Vol. 1. Cim. Petroleum Soc/SPE International Technical Management, Calgary, Canada, 10–13 June 1990.

Crowe, C.W., 1991a. Laboratory study provides guidelines for selecting clay stabilizers. SPE Unsolicited Pap SPE-21556, Dowell Schlumberger.

Crowe, C.W., 1991b. Laboratory study provides guidelines for selecting clay stabilizers. In: Proceedings Volume. SPE Oilfield Chemical International Symposium, Anaheim, California, 20–22 February 1991, pp. 499–504.

Doleschall, S., Milley, G., Paal, T., 1987. Control of clays in fluid reservoirs. In: Proceedings Volume. 4th BASF AG et al Enhanced Oil Recovery Europe Symp. (Hamburg, Ger, 27–28 October 1987), pp. 803–812.

Durand, C., Onaisi, A., Audibert, A., Forsans, T., Ruffet, C., 1995a. Influence of clays on borehole stability: A literature survey: Pt.1: Occurrence of drilling problems physico-chemical description of clays and of their interaction with fluids. Rev. Inst. Franc. Pet. 50 (2), 187–218.

Durand, C., Onaisi, A., Audibert, A., Forsans, T., Ruffet, C., 1995b. Influence of clays on borehole stability: A literature survey: Pt.2: Mechanical description and modelling of clays and shales drilling practices versus laboratory simulations. Rev. Inst. Franc. Pet. 50 (3), 353–369.

Eoff, L.S., Reddy, B.R., Wilson, J.M., 2006. Compositions for and methods of stabilizing subterranean formations containing clays. US Patent 7 091 159, assigned to Halliburton Energy Services, Inc., Duncan, OK, 15 August 2006. <http://www.freepatentsonline.com/7091159.html>.

Evans, B., Ali, S., 1997. Selecting brines and clay stabilizers to prevent formation damage. World Oil 218 (5), 65–68.

Gazaniol, D., Forsans, T., Boisson, M.J.F., Piau, J.M., 1994. Wellbore failure mechanisms in shales: Prediction and prevention. In: Proceedings Volume. Vol. 1. SPE Europe Petroleum Conference, London, UK, 25–27 October 1994, pp. 459–471.

Gazaniol, D., Forsans, T., Boisson, M.J.F., Piau, J.M., 1995. Wellbore failure mechanisms in shales: Prediction and prevention. J. Pet. Technol. 47 (7), 589–595.

Grim, R.E., 1968. Clay Mineralogy, second ed. McGraw-Hill, New York.

Hale, A.H., van Oort, E., 1997. Efficiency of ethoxylated/propoxylated polyols with other additives to remove water from shale. US Patent 5 602 082, 11 February 1997.

Hall, B.E., Szememyei, C.A., 1992. Fluid additive and method for treatment of subterranean formations. US Patent 5 089 151, assigned to Western Co. North America, 18 February 1992.

Halliday, W.S., Thielen, V.M., 1987. Drilling mud additive. US Patent 4 664 818, assigned to Newpark Drilling Fluid In, 12 May 1987.

Himes, R.E., 1992. Method for clay stabilization with quaternary amines. US Patent 5 097 904, assigned to Halliburton Co., 24 March 1992.

Himes, R.E., Vinson, E.F., 1989. Fluid additive and method for treatment of subterranean formations. US Patent 4 842 073, 27 June 1989.

Himes, R.E., Vinson, E.F., 1991. Environmentally safe salt replacement for fracturing fluids. In: Proceedings Volume. SPE East Reg Conference, Lexington, KY, 23–25 October 1991, pp. 237–248.

Himes, R.E., Vinson, E.F., Simon, D.E., 1989. Clay stabilization in low-permeability formations. In: Proceedings Volume. SPE Production Operation Symposium, Oklahoma City, 12–14 March 1989, pp. 507–516.

Himes, R.E., Parker, M.A., Schmelzl, E.G., 1990. Environmentally safe temporary clay stabilizer for use in well service fluids. In: Proceedings Volume. Vol. 3. Cim Petroleum Soc/SPE International Technical Management, Calgary, Canada, 10–13 June 1990.

Hoffmann, R., Lipscomb, W.N., 1962. Theory of polyhedral molecules. I. Physical factorizations of the secular equation. J. Chem. Phys. 36 (8), 2179–2189.

Huber, J., Plank, J., Heidlas, J., Keilhofer, G., Lange, P., 2009. Additive for drilling fluids. US Patent 7 576 039, assigned to BASF Construction Polymers GmbH, Trostberg, DE, 18 August 2009. <http://www.freepatentsonline.com/7576039.html>.

Kippie, D.P., Gatlin, L.W., 2009. Shale inhibition additive for oil/gas down hole fluids and methods for making and using same. US Patent 7 566 686, assigned to Clearwater International, LLC, Houston, TX, 28 July 2009. <http://www.freepatentsonline.com/7566686.html>.

Klein, H.P., Godinich, C.E., 2006. Drilling fluids. US Patent 7 012 043, assigned to Huntsman Petrochemical Corporation, The Woodlands, TX, March 14 2006. <http://www.freepatentsonline.com/7012043.html>.

Mody, F.K., Hale, A.H., 1993. A borehole stability model to couple the mechanics and chemistry of drilling fluid shale interaction. In: Proceedings Volume. SPE/IADC Drilling Conference, Amsterdam, Netherland 23–25 February 1993, pp. 473–490.

Mooney, R.W., Keenan, A.G., Wood, L.A., 1952. Adsorption of water vapor by montmorillonite. II. Effect of exchangeable ions and lattice swelling as measured by X-ray diffraction. J. Am. Chem. Soc. 74 (6), 1371–1374.

Murphy, C.B., Fabri, J.O., Reilly Jr., P.B., 2012. Treatment of subterranean formations. US Patent 8 157 010, assigned to Polymer Ventures, Inc., Charleston, SC, 17 April 2012. <http://www.freepatentsonline.com/8157010.html>.

Murray, H.H., 2007. Applied Clay Mineralogy: Occurrences, Processing, and Application of Kaolins, Bentonites, Palygorskite-Sepiolite, and Common Clays, vol 2. Elsevier, Amsterdam.

Norrish, K., 1954. The swelling of montmorillonite. Discuss. Faraday Soc. 18, 120–134.

Palumbo, S., Giacca, D., Ferrari, M., Pirovano, P., 1989. The development of potassium cellulosic polymers and their contribution to the inhibition of hydratable clays. In: Proceedings Volume. SPE Oilfield Chemical International Symposium, Houston, 8–10 February 1989, pp. 173–182.

Patel, A.D., McLaurine, H.C., 1993. Drilling fluid additive and method for inhibiting hydration. CA Patent 2 088 344, assigned to M I Drilling Fluids Co., 11 October 1993.

Patel, M.A., Patel, H.S., 2012. A review on effects of stabilizing agents for stabilization of weak soil. Civil and Environmental Research 2 (6), 1–7.

Patel, A.D., Stamatakis, E., Davis, E., Friedheim, J., 2007. High performance water based drilling fluids and method of use. US Patent 7 250 390, assigned to M-I L.L.C., Houston, TX, 31 July 2007. <http://www.freepatentsonline.com/7250390.html>.

Poelker, D.J., McMahon, J., Schield, J.A., 2009. Polyamine salts as clay stabilizing agents. US Patent 7 601 675, assigned to Baker Hughes Incorp., Houston, TX, 13 October 2009. <http://www.freepatentsonline.com/7601675.html>.

Scheuerman, R.F., Bergersen, B.M., 1989. Injection water salinity, formation pretreatment, and well operations fluid selection guidelines. In: Proceedings Volume. SPE Oilfield Chemical International Symposium, Houston, 8–10 February 1989, pp. 33–49.

Schield, J.A., Naiman, M.I., Scherubel, G.A., 1991. Polyimide quaternary salts as clay stabilization agents. GB Patent 2 244 270, assigned to Petrolite Corp., 27 November 1991.

Smith, K.W., 2009. Well drilling fluids. US Patent 7 576 038, assigned to Clearwater International, L.L.C., Houston, TX, August 18 2009. <http://www.freepatentsonline.com/7576038.html>.

Smith, C.K., Balson, T.G., 2000. Shale-stabilizing additives. GB Patent 2 340 521, assigned to Sofitech NV and Dow Chemical Co., 23 February 2000.

Smith, C.K., Balson, T.G., 2004. Shale-stabilizing additives. US Patent 6 706 667, 16 March 2004. <http://www.freepatentsonline.com/6706667.html>.

Stowe, C., Bland, R.G., Clapper, D., Xiang, T., Benaissa, S., 2002. Water-based drilling fluids using latex additives. GB Patent 2 363 622, assigned to Baker Hughes Inc., 2 January 2002.

Suratman, I., 1985. A study of the laws of variation (kinetics) and the stabilization of swelling of clay (contribution a l'etude de la cinetique et de la stabilisation du gonflement des argiles). Ph.D. Thesis, Malaysia.

Tan, C.P., Richards, B.G., Rahman, S.S., Andika, R., 1997. Effects of swelling and hydrational stress in shales on wellbore stability. In: Proceedings Volume. SPE Asia Pacific Oil & Gas Conference, Kuala Lumpur, Malaysia, 14–16 April 1997, pp. 345–349.

Thomas, T.R., Smith, K.W., 1993. Method of maintaining subterranean formation permeability and inhibiting clay swelling. US Patent 5 211 239, assigned to Clearwater Inc., 18 May 1993.

Tshibangu, J.P., Sarda, J.P., Audibert-Hayet, A., 1996. A study of the mechanical and physicochemical interactions between the clay materials and the drilling fluids: Application to the boom clay (Belgium) (etude des interactions mecaniques et physicochimiques entre les argiles et les fluides de forage: Application a l'argile de boom (Belgique)). Rev. Inst. Franc. Pet. 51 (4), 497–526.

Van Oort, E., 1997. Physico-chemical stabilization of shales. In: Proceedings Volume. SPE Oilfield Chemical International Symposium, Houston, 18–21 February 1997, pp. 523–538.

Waldmann, J.J., 1997. Agents having high nitrogen content and high cationic charge based on dicyanimide dicyandiamide or guanidine and inorganic ammonium salts. US Patent 5 659 011, 19 August 1997. <http://www.freepatentsonline.com/5659011.html>.

Westerkamp, A., Wegner, C., Mueller, H.P., 1991. Borehole treatment fluids with clay swelling-inhibiting properties (ii) (bohrloch- behandlungsfluessigkeiten mit tonquellungsinhibierenden eigenschaften (ii)). EP Patent 451 586, assigned to Bayer AG, 16 October 1991.

Whitfill, D.L., Pober, K.W., Carlson, T.R., Tare, U.A., Fisk, J.V., Billingsley, J.L., 2005. Method for drilling depleted sands with minimal drilling fluid loss. US Patent 6 889 780, assigned to Halliburton Energy Services, Inc., Duncan, OK, 10 May 2005. <http://www.freepatentsonline.com/6889780.html>.

Wilcox, R.D., 1990. Surface area approach key to borehole stability. Oil Gas J. 88 (9), 66–80.

Yeager, R.R., Bailey, D.E., 1988. Diesel-based gel concentrate improves rocky mountain region fracture treatments. In: Proceedings Volume. SPE Rocky Mountain Regional Management, Casper, Wyo, 11–13 May 1988, pp. 493–497.

Zhou, Z.J., Gunter, W.D., Jonasson, R.G., 1995. Controlling formation damage using clay stabilizers: A review. In: Proceedings Volume-2. No. CIM 95–71. 46th Annual Cim. Petroleum Society Techical Management, Banff, Canada 14–17 May 1995.

Chapter 9

pH Control Additives

The pH means *potentia hydrogenii* and is defined as the negative decadic logarithm of the concentration of protons or hydrogenated protons, respectively. For example, a concentration of protons of 10^{-12} mol l^{-1} means a pH of 12.

Similar considerations are valid for equilibrium constants K, that are sometimes expressed in terms of pK values.

Water dissociates to a small extent into protons or more correctly into hydrogenated protons and hydroxyl ions, i.e.,

$$H_2O \rightarrow H^+ + OH^-. \tag{9.1}$$

In the presence of water, the proton as such is not stable, but will be attached to another molecule of water:

$$H^+ + H_2O \rightarrow H_3O^+. \tag{9.2}$$

For simplicity we will deal often only with H^+. The equilibrium constant of the dissociation reaction, Eq. (9.1), is

$$K_{H_2O} = \frac{[H^+][OH^-]}{[H_2O]}. \tag{9.3}$$

The equilibrium constant K_{H_2O} has in general a physical dimension. Since the concentration has the physical dimension of mol l^{-1}, K_{H_2O} has the physical dimension of mol l^{-1}.

THEORY OF BUFFERS

The theory of buffers is a common part of physical chemistry, e.g., in the book of Chang (2000). A protic acid A = B–H decomposes in the first step into a proton H^+ and into a base B^-.

$$\begin{aligned} BH &\rightarrow H^+ + B^-, \\ B^- + H_2O &\rightarrow BH + OH^-. \end{aligned} \tag{9.4}$$

Hydraulic Fracturing Chemicals and Fluids Technology. http://dx.doi.org/10.1016/B978-0-12-411491-3.00009-1
© 2013 Elsevier Inc. All rights reserved.

For the two reactions in Eq. (9.4) we have two equilibrium constants, for the acid

$$K_A = \frac{[H^+][B^-]}{[BH]} \tag{9.5}$$

and for the base

$$K_B = \frac{[OH^-][BH]}{[B^-][H_2O]}. \tag{9.6}$$

Summing up the two individual equations of Eq. (9.4), yields again the dissociation equation if water.

$$H_2O \rightarrow H^+ + OH^-. \tag{9.7}$$

Therefore,

$$K_A K_B = K_{H_2O}. \tag{9.8}$$

As a further equation we have the balance of water

$$[H_2O]_0 = [H_2O] + [H^+]. \tag{9.9}$$

Here $[H_2O]_0$ is the concentration of the undissociated water. From the principle of electroneutrality it follows

$$[H^+] = [OH^-]. \tag{9.10}$$

The equilibrium constant for water is very small, therefore we may safely approximate $[H_2O]_0 \approx [H_2O]$.

The relation

$$[H_2O]_0 K_{H_2O} = [H^+][OH^-] \tag{9.11}$$

is addressed as the ionic product of water. It is close to $10^{-14}\ \mathrm{mol}^2\ \mathrm{l}^{-2}$.

Inserting Eqs. (9.10) and (9.9) into Eq. (9.3) results in

$$K_{H_2O} = \frac{[H^+]^2}{[H_2O]_0 - [H^+]}. \tag{9.12}$$

From the relations Eq. (9.11) or Eq. (9.9) the pH of pure water can be calculated.

An extension of the above equations to weak acids and bases, Eq. (9.4) will allow to calculate the pH value in buffer systems, as well as titration curves. Common aqueous buffer ingredients are shown in Table 9.1 and in Figures 9.1 and 9.2.

Buffers, necessary to adjust and maintain the pH, can be salts of a weak acid and a weak base. Examples are carbonates, bicarbonates, and hydrogen phosphates, such as formic acid, fumaric acid, and sulfamic acid (Nimerick, 1996).

Increased temperature stability of various gums can be achieved by adding sodium bicarbonate to the fracturing fluid and thus raising its pH to 9.2–10.4.

TABLE 9.1 Common Buffer Solutions (Kolditz, 1967)

Reaction	pK_A
$HCl \rightarrow H^+ + Cl^-$	−7
$H_2SO_4 \rightarrow H^+ + HSO_4^-$	−3
$HSO_4^- \rightarrow SO_4^{2-} + H^+$	1.92
$H_2SO_3 \rightarrow HSO_3^- + H^+$	1.92
$HF \rightarrow F^- + H^+$	3.14
$CH_3COOH \rightarrow CH_3COO^- + H^+$	4.75
$H_2S \rightarrow HS^- + H^+$	6.92
$NH_4^+ \rightarrow NH_3 + H^+$	9.25
$H_2O \rightarrow H^+ + OH^-$	15.47
Oxalic acid/hydrogen oxalate	1.27
Maleic acid/hydrogen maleate	1.92
Fumaric acid/hydrogen fumarate	3.03
Citric acid/hydrogen citrate	3.13
Sulfamic acid/sulfamate	1.0
Formic acid/formate	3.8
Acetic acid/acetate	4.7
Dihydrogen phosphate/hydrogen phosphate	7.1
Ammonium/ammonia	9.3
Bicarbonate/carbonate	10.4
Fumaric acid/hydrogen fumarate	3.0
Benzoic acid/benzoate	4.2

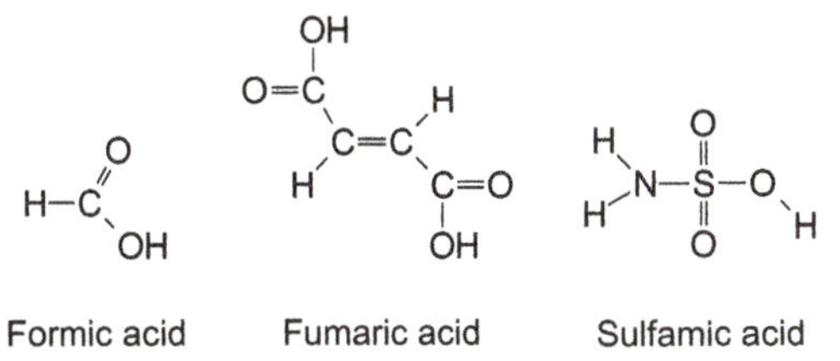

Formic acid Fumaric acid Sulfamic acid

FIGURE 9.1 Weak organic acids.

pH CONTROL

The buffer to be used depends on the intended pH range. For fracturing fluids, several buffer systems have been recommended that are summarized in Table 9.2.

For applications with zirconium crosslinked guar gums, the use of a combination of buffers together with the use of an α-hydroxycarboxylic acid

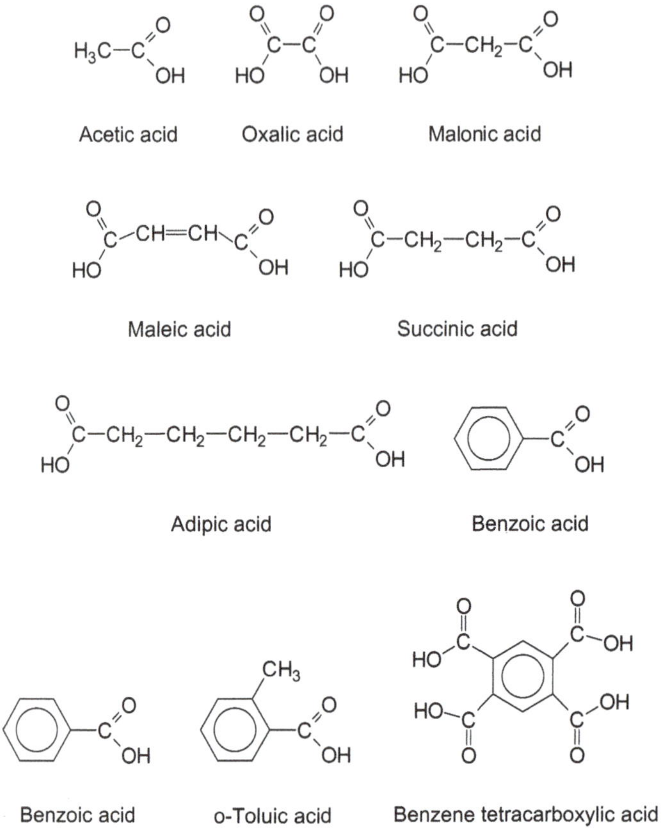

FIGURE 9.2 Carbonic and dicarbonic acids.

or its salt offers an advantage in that the crosslink delay time of the newly proposed fluid is controllable over a wider pH range (Moorhouse and Matthews, 2004).

TABLE 9.2 Buffer Systems (Putzig, 2012)

pH Range	Buffer System
5–7	Fumaric acid
5–7	Sodium diacetate
7–8.5	Sodium bicarbonate
9–12	Sodium carbonate
9–12	Sodium hydroxide

REFERENCES

Chang, R., 2000. Physical chemistry for the chemical and biological sciences. University Science Books, Sausalito, California (Chapter 11).

Kolditz, L. (Ed.), 1967. Anorganikum: Lehr- und Praktikumsbuch der anorganischen Chemie; mit einer Einführung in die physikalische Chemie. Dt. Verl. der Wiss., Berlin, p. 413.

Moorhouse, R., Matthews, L.E., 2004. Aqueous based zirconium (iv) crosslinked guar fracturing fluid and a method of making and use therefor. US Patent 6 737 386, assigned to Benchmark Research and Technology Inc., Midland, TX, 18 May 2004. <http://www.freepatentsonline.com/6737386.html>.

Nimerick, K., 1996. Fracturing fluid and method. GB Patent 2 291 907, assigned to Sofitech NV, 7 February 1996.

Putzig, D.E., 2012. Zirconium-based cross-linking composition for use with high pH polymer solutions. US Patent 8 247 356, assigned to Dorf Ketal Speciality Catalysts, LLC, Stafford, TX, 21 August 2012. <http://www.freepatentsonline.com/8247356.html>.

Chapter 10

Surfactants

Surface active agents are included in most aqueous treating fluids to improve the compatibility of aqueous fluids with the hydrocarbon-containing reservoir. To achieve a maximal conductivity of hydrocarbons from subterranean formations after fracture or other stimulation, it is the practice to cause the formation surfaces to be water-wet.

Alkylamino phosphonic acids and fluorinated alkylamino phosphonic acids adsorb onto solid surfaces, particularly onto surfaces of carbonate materials in subterranean hydrocarbon-containing formations, in a very thin layer. The layer is only one molecule thick and thus significantly thinner than a layer of water or a water-surfactant mixture on water-wetted surfaces (Penny, 1987a,b; Penny and Briscoe, 1987).

These compounds so adsorbed resist or substantially reduce the wetting of the surfaces by water and hydrocarbons and provide high interfacial tensions between the surfaces and water and hydrocarbons. The hydrocarbons displace injected water, leaving a lower water saturation and an increased flow of hydrocarbons through capillaries and flow channels in the formation.

PERFORMANCE STUDIES

The primary purpose of surfactants used in stimulating sandstone reservoirs is to reduce the surface tension and the contact angle thus to provide control of the fluid loss. However, many of the surfactants are adsorbed rapidly within the first few inches of the sandstone formations. In this way, their effectiveness with respect to deeper penetration is reduced.

Experimental and field studies of various surfactants used in the oilfield have been described (Paktinat et al., 2007).

Several different surfactants were investigated to determine their adsorption properties when injected into a laboratory sandpacked column. In addition, field data were collected from Bradford, Balltown, and Speechley sandstone formations. The correlation between laboratory and field data was confirmed.

Reservoirs that were treated with microemulsion fluids demonstrate exceptional water recoveries when compared with conventional surfactant

Hydraulic Fracturing Chemicals and Fluids Technology. http://dx.doi.org/10.1016/B978-0-12-411491-3.00010-8
© 2013 Elsevier Inc. All rights reserved.

treatments. These investigations are considered to serve as case studies and can be used to minimize formation damage (Paktinat et al., 2007).

VISCOELASTIC SURFACTANTS

Typical viscoelastic surfactants are *N*-erucyl-*N*,*N*-bis(2-hydroxyethyl)-*N*-methyl ammonium chloride and potassium oleate, solutions of which form gels when mixed with corresponding activators such as sodium salicylate and potassium chloride (Jones and Tustin, 2007).

A methyl quaternized erucyl amine is useful for aqueous viscoelastic surfactant-based fracturing fluids in high-temperature and high-permeability formations (Gadberry et al., 1999).

A problem associated with the use of viscoelastic surfactants is that stable oil-in-water emulsions are often formed between the low-viscosity surfactant solution, i.e., the broken gel and the reservoir hydrocarbons. As a consequence, a clean separation of the two phases may be difficult to achieve, complicating the cleanup of wellbore fluids. Such emulsions are believed to form because conventional wellbore fluid viscoelastic surfactants have little or no solubility in organic solvents (Jones and Tustin, 2007).

Cationic Surfactants

A number of cationic surfactants, based on quaternary ammonium and phosphonium salts, are known to exhibit solubility in water and hydrocarbons and are frequently as such used as phase-transfer catalysts (Starks et al., 1994).

However, the particular cationic surfactants which form viscoelastic solutions in aqueous media are poorly soluble in hydrocarbons, and are characterized by partition coefficients $K_{o,w}$ for a surfactant in oil and water close to zero.

The partition coefficient of a substance is the ratio of the concentrations in equilibrium in two non-miscible fluids, such as oil and water.

$$K_{o,w} = \frac{c_o}{c_w}. \tag{10.1}$$

The partition coefficient can be determined by various analytical techniques (Sharaf et al., 1986).

For example, cyclic voltammetry has been used for the determination of the critical micelle concentration of surfactants, self-diffusion coefficient of micelles, and the partition coefficient (Mandal and Nair, 1991). Also, high-performance liquid chromatography is a suitable technique (Terweij-Groen et al., 1978).

Typically, the high solubility of a cationic surfactant in hydrocarbon solvents is promoted by multiple long-chain alkyl groups attached to a head group, as can be found in hexadecyltributylphosphonium ions and trioctylmethylammonium ions.

In contrast, cationic surfactants which form viscoelastic solutions generally have only one long straight hydrocarbon chain per surfactant head moiety (Jones and Tustin, 2007).

Anionic Surfactants

A few anionic surfactants exhibit a high solubility in hydrocarbons but low solubility in aqueous solutions. A well-known example is sodium bis (2-ethylhexyl) sulfosuccinate (Manoj et al., 1996). This compound does not form viscoelastic solutions in aqueous media. So, the addition of a salt causes precipitation. Thermodynamic studies suggest that the micellization process is endothermic in nature so that it is mainly an entropy governed process.

The solubility of a surfactant in hydrocarbon tends to increase as the size of the side chain decreases. It is believed that this occurs because smaller side chains cause less disruption to the formation of inverse micelles by the surfactant in the hydrocarbon, such inverse micelles promoting solubility in the hydrocarbon (Jones and Tustin, 2007).

By altering the degree and type of branching from the principal straight chain, the surfactant can be tailored to be more or less soluble in a particular hydrocarbon. Preferably the side chain is bonded to the α-carbon atom. By locating the side chain close to the charged head group promotes the most favorable combinations of viscoelastic and solute properties. The synthesis of a β-branched fatty acid is shown schematically in Figure 10.1.

Preparation 10.1. Synthesis of 2-methyl methyl oleate (Jones and Tustin, 2007): Sodium hydride is washed with heptane and then suspended in tetrahydrofuran. 1,3-Dimethyl-3,4,5,6-tetrahydro-2(1H)-pyrimidinone is then added and the mixture is stirred in a nitrogen atmosphere. Then methyl oleate

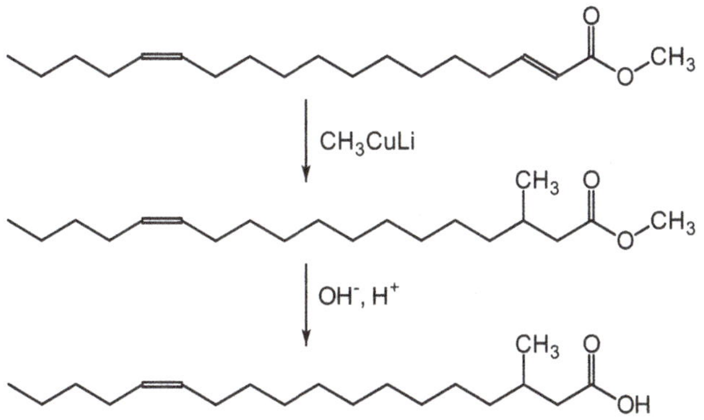

CH_3CuLi

OH^-, H^+

FIGURE 10.1 Synthesis of β-branched fatty acids (Jones and Tustin, 2007).

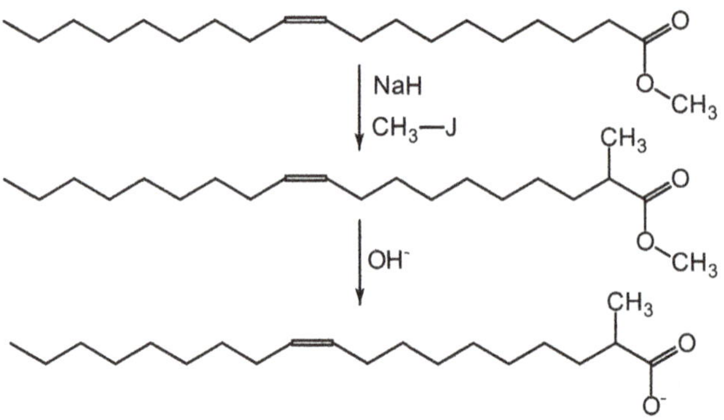

FIGURE 10.2 Synthesis of 2-methyl methyl oleate (Jones and Tustin, 2007).

is dropwise added over a period of 2h and the resulting mixture is heated to reflux for 12h and then cooled to 0°C.

Further, methyl iodide is added dropwise and the reaction mixture is again heated to reflux for a further 2h. Afterwards the reaction mixture is cooled to 0°C and quenched with water, concentrated in vacuo and purified by column chromatography to give the end product 2-methyl methyl oleate as a yellow oil. ■

The synthesis of 2-methyl methyl oleate and the subsequent hydrolysis of the ester is shown in Figure 10.2. A rigid gel is formed when a 10% solution of potassium 2-methyl oleate is mixed with an equal volume of a brine containing 16% KCl.

Contacting this gel with a representative hydrocarbon, such as heptane, results in a dramatic loss of the viscosity and the formation of two low-viscosity clear solutions:

1. An upper oil phase.
2. A lower aqueous phase.

The formation of an emulsion was not observed. Thin-layer chromatography and infrared spectroscopy showed the presence of the branched oleate in both phases.

The gel is apparently broken by a combination of micellar rearrangement and dissolution of the branched oleate in the oil phase. Consequently the breaking rate of the branched oleate is faster than that of the equivalent linear oleate. This is demonstrated in Figure 10.3 which is a graph of gel strength against time at room temperature for an unbranched potassium oleate gel and a branched potassium 2-methyl oleate gel.

Both gels were prepared from 10% solutions of the respective oleate mixed with equal volumes of a brine containing 16% KCl. Each gel was then contacted with an equal volume of heptane.

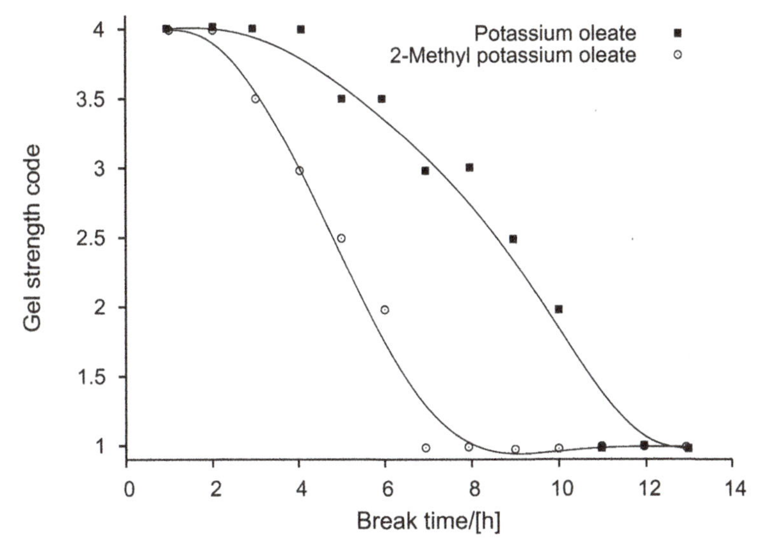

FIGURE 10.3 Gel strength versus time (Jones and Tustin, 2007).

TABLE 10.1 Gel Strength Codings (Jones and Tustin, 2010)

Number	Description
1	Original viscosity
2	Weak flowing gel
3	Tonguing gel
4	Deformable nonflowing gel

The gel strength is a semiquantitative measure of the flowability of the surfactant-based gel relative to the flowability of the precursor fluid before addition of the surfactant. There are four gel strength codings. These codings are summarized in Table 10.1.

Using infrared spectroscopy, the value of K_{ow} for the potassium 2-methyl oleate of the broken branched gel was measured as 0.11. In contrast the value of K_{ow} for the potassium oleate of the broken unbranched gel was measured as effectively zero.

The rapid breakdown of the branched oleate surfactant gels, with little or no subsequent emulsion, leads to the expectation that these gels will be particularly suitable for use as wellbore fluids, such as fluids for hydraulic fracturing of oil-bearing zones. Excellent cleanup of the fluids and reduced impairment of zone matrix permeability can also be expected because emulsion formation can be avoided (Jones and Tustin, 2007).

Anionic Brominated Surfactants

In the case of ionic surfactants, an excess of counterions reduces the repulsive interaction between the charged head groups. The surfactant molecules at both ends of such cylindrical aggregates bear an excess of energy in comparison to the molecules in the inner of the cylindrical part. This excess energy is addressed as end cap energy and it is the driving force for the linear growth of cylindrical micelles. Subsequently, the micelles entangle with each other, which results in a viscoelastic behavior (Lee et al., 2010).

A problem associated with the use of such surfactants is the potential formation of stable oil in water emulsions during a cleanup operation. This behavior arises due to the limited solubility in hydrocarbon of conventional viscoelastic surfactants.

However, it has been demonstrated that a conventional surfactant, such as potassium oleate, can be brominated and in this way the properties are improved (Lee et al., 2010). The bromination of oleic acid is done in a hexane solution to which a solution of HBr in acetic acid is added. In this way 9-bromo stearate is obtained.

The partition coefficient and the gel break time for various concentrations of 9-bromo stearate and potassium oleate in 8% KCl solution are shown in Table 10.2. As can be seen from Table 10.2, the bromination causes a significant change in the partition coefficient, but does not change the gel break time.

Furthermore, the bromination of the hydrocarbon chain keeps the viscoelasticity of the surfactant. Thus, the essential property for oilfield applications is maintained. When the shear stress is removed in the formation, the solution reverts to its viscous state. However, the bromination reduces the zero shear viscosity from 528 Pa s to 180 Pa s. In contrast, the shear viscosity at 100 s^{-1} is reduced from 0.48 Pa s to 0.16 Pa s by the bromination at a concentration of 5% in 8% KCl.

TABLE 10.2 Partition Coefficient and Gel Break Time (Lee et al., 2010)

Concentration (%)	Partition (%)		Gel Break Time (h)	
	BST	OLE	BST	OLE
1	17.1	0	12	14
5	13.28	0	82	84
10	13.49	0	94	98

BST 9-Bromo stearate and OLE Oleate.

REFERENCES

Gadberry, J.F., Hoey, M.D., Franklin, R., Del Carmen Vale, G., Mozayeni, F., 1999. Surfactants for hydraulic fracturing compositions. US Patent 5979555, assigned to Akzo Nobel NV, 9 November 1999.

Jones, T.G.J., Tustin, G.J., 2007. Surfactant comprising alkali metal salt of 2-methyl oleic acid or 2-ethyl oleic acid. US Patent 7196041, assigned to Schlumberger Technology Corp., Ridgefield, CT, 27 March 2007. <http://www.freepatentsonline.com/7196041.html>.

Jones, T.G.J., Tustin, G.J., 2010. Process of hydraulic fracturing using a viscoelastic wellbore fluid. US Patent 7655604, assigned to Schlumberger Technology Corp., Ridgefield, CT, 2 February 2010. <http://www.freepatentsonline.com/7655604.html>.

Lee, L., Salimon, J., Yarmo, M.A., Misran, M., 2010. Viscoelastic properties of anionic brominated surfactants. Sains Malays. 39 (5), 753–760.

Mandal, A.B., Nair, B.U., 1991. Cyclic voltammetric technique for the determination of the critical micelle concentration of surfactants, self-diffusion coefficient of micelles, and partition coefficient of an electrochemical probe. J. Phys. Chem. 95 (22), 9008–9013. http://dx.doi.org/10.1021/j100175a106.

Manoj, K.M., Jayakumar, R., Rakshit, S.K., 1996. Physicochemical studies on reverse micelles of sodium bis(2-ethylhexyl) sulfosuccinate at low water content. Langmuir 12 (17), 4068–4072. http://dx.doi.org/10.1021/la950279a.

Paktinat, J., Pinkhouse, J.A., Williams, C., Clark, G.A., Penny, G.S., 2007. Field case studies: damage preventions through leakoff control of fracturing fluids in marginal/low-pressure gas reservoirs. SPE Prod. Oper. 22 (3), 357–367.

Penny, G.S., 1987a. Method of increasing hydrocarbon production from subterranean formations. US Patent 4 702 849, 27 October 1987.

Penny, G.S., 1987b. Method of increasing hydrocarbon productions from subterranean formations. EP Patent 234910, 2 September 1987.

Penny, G.S., Briscoe, J.E., 1987. Method of increasing hydrocarbon production by remedial well treatment. CA Patent 1216 416, 13 January 1987.

Sharaf, M.A., Illman, D.L., Kowalski, B.R., 1986. Chemometrics. Wiley, New York.

Starks, C.M., Liotta, C.L., Halpern, M., 1994. Phase-transfer catalysis: fundamentals, applications, and industrial perspectives. Chapman & Hall, New York.

Terweij-Groen, C.P., Heemstra, S., Kraak, J.C., 1978. Distribution mechanism of ionizable substances in dynamic anion-exchange systems using cationic surfactants in high-performance liquid chromatography. J. Chromatogr. A 161, 69–82. http://dx.doi.org/10.1016/S0021-9673(01)85213-4.

Chapter 11

Scale Inhibitors

In certain operations in petroleum industries, such as production, stimulation, and transport, there is some risk of scale deposition. Scaling can occur when a solution becomes supersaturated, which occurs mostly if the temperature changes in the course of injection operations.

Also, if two chemicals that will form a precipitate are brought together, a scale is formed, e.g., if a hydrogen fluoride solution meets calcium ions. From a thermodynamic perspective, there is a stable region, a metastable region, and an unstable region, separated by the binodal curve and the spinodale curve, respectively.

Scales may consist of calcium carbonate, barium sulfate, gypsum, strontium sulfate, iron carbonate, iron oxides, iron sulfides, and magnesium salts (Keatch, 1998). There are monographs, e.g., *Corrosion and Scale Handbook* (Becker, 1998), as well as reviews (Crabtree et al., 1999) on scale depositions available in the literature. Case studies have been presented for North Sea carbonate reservoirs (Jordan et al., (2003,2005)) and Gulf of Mexico (Jordan et al., 2002). A more recent topic focuses on green systems (Frenier and Hill, 2004; Hasson et al., 2011).

CLASSIFICATION AND MECHANISM

The problem is basically similar to preventing scale inhibition in washing machines. Therefore, similar chemicals are used to prevent scale deposition. Scale inhibition can be achieved either by adding substances that react with potential scale-forming substances so that from the view of thermodynamics the stable region is reached or by adding substances that suppress crystal growth.

Conventional scale inhibitors are hydrophilic, i.e., they dissolve in water. In the case of downhole squeezing, it is desirable that the scale inhibitor is adsorbed on the rock to avoid washing out the chemical before it can act as desired. However, adsorption on the rock may change the surface tension and the wettability of the system. To overcome these disadvantages, oil-soluble scale inhibitors have been developed. Coated inhibitors are also available.

Frequently, scale inhibitors are not applied as such, but in combination with corrosion inhibitors (Martin et al., 2005). Scale inhibitors can be classified into two main groups, namely

Hydraulic Fracturing Chemicals and Fluids Technology. http://dx.doi.org/10.1016/B978-0-12-411491-3.00011-X
© 2013 Elsevier Inc. All rights reserved.

- Thermodynamic inhibitors.
- Kinetic inhibitors.

Scale prevention is important to ensure continuous production from existing reserves that produce brine. Wells can be abandoned prematurely because of poor management of scale and corrosion (Kan and Tomson, 2010). Two ways by which the kinetic scale inhibitor operates are known, by Viloria et al. (2010):

1. Adsorption effects.
2. Morphologic changes of the growing sites.

Due to adsorption effects, the inhibitor molecules occupy the nucleation sites which are preferred by the scale-forming molecules. Thus, crystals cannot find active places to adhere to the surface and, therefore, crystal nucleation is not promoted.

Another inhibitor mechanism is based on the mechanism of adsorption, i.e., morphologic changes can prevent the formation of crystals in the presence of the inhibitor. Depending on the inhibitor characteristics and the nature of the substrate, it is possible that the inhibitor will be adsorbed over the crystalline net, forming complex surfaces or nets which have difficulty remaining and growing in active places.

Sea water often reacts with the formation water in offshore fields to produce barium, calcium, and strontium sulfate deposits that hinder oil production. In some fields, $CaCO_3$ is a major problem.

In some regions, the formation water chemistry varies considerably (Duccini et al., 1997). For example, in the Central North Sea Province, the levels of barium ions vary from a few $mg\,l^{-1}$ to $g\,l^{-1}$. Further, the pH varies from 4.4 to 7.5. Ultimately, a pH of 11.7 has been measured. In the southern region of the North Sea, the waters have a high salinity and are rich in sulfate and acidic compounds. The ideal scale inhibitor should have the following properties (Duccini et al., 1997):

- Effective scale control at low inhibitor concentration.
- Compatibility with sea and formation water.
- Balanced adsorption-desorption properties allowing the chemicals to be slowly and homogeneously released into the production water.
- High thermal stability.
- Low toxicity and high biodegradability.
- Low cost.

Scale inhibitors are coarsely classified as organic and inorganic (Viloria et al., 2010). The inorganic types include condensed phosphate, such as poly(metaphosphate)s or phosphate salts. Suitable organic scale inhibitors available are poly(acrylic acid) (PAA), phosphinocarboxylic acid, sulfonated polymers, and phosphonates (Duccini et al., 1997).

TABLE 11.1 Types of Scale Inhibitors (Viloria et al., 2010)

Inhibitor Type	Limitations
Inorganic poly(phosphate)s	Suffer hydrolysis and can precipitate as calcium phosphates because of temperature, pH, solution quality, concentration, phosphate type and the presence of some enzymes
Organic poly(phosphate)s	Suffer hydrolysis with temperature. Not effective at high calcium concentrations. Must be applied in high doses
Polymers based on carboxylic acids	Limited calcium tolerance (2000 ppm) although some can work at concentrations higher than 5000 ppm Larger concentrations are needed
Ethylenediamine tetraacetic acid	Expensive

Phosphonates are maximally effective at high temperatures whereas sulfonated polymers are maximally effective at low temperatures (Talbot et al., 2009). Copolymers that contain both phosphonate and sulfonate moieties can produce and enhanced scale inhibition over a range of temperatures. A phosphonate end-capped vinyl sulfonic acid/acrylic acid copolymer has been shown to be particularly useful in the scale inhibition of barium sulfate scale in water-based systems (Talbot et al., 2009). The basic issues of scale inhibitors are given in Table 11.1.

Thermodynamic Inhibitors

Thermodynamic inhibitors are complexing and chelating agents, suitable for specific scales. For example, for scale inhibition of barium sulfate, common chemicals are ethylenediamine tetraacetic acid (EDTA) and nitrilotriacetic acid. The solubility of calcium carbonate can be influenced by varying the pH or the partial pressure of carbon dioxide (CO_2). The solubility increases with decreasing pH and increasing partial pressure of CO_2, and it decreases with temperature.

However, usually the solubility increases with higher temperature. The temperature coefficient of solubility is dependent on the enthalpy of dissolution. An exothermic enthalpy of dissolution causes a decrease in solubility with increased temperature, and vice versa.

Kinetic Inhibitors

Kinetic inhibitors for hydrate formation may also be effective in preventing scale deposition (Sikes and Wierzbicki, 1996). This may be understood in terms of stereospecific and nonspecific mechanisms of scale inhibition.

Adherence Inhibitors

Another mechanism of scale inhibition is based on adherence inhibitors. Some chemicals simply suppress the adherence of crystals to the metal surfaces. These are surface active agents.

Interference of Chelate Formers

Trace amounts of metal chelate-forming additives, which are used in fracture fluids, have been shown to have a debilitating effect on the performance of widely used barium sulfate scale inhibitors. Ethylenediamine tetraacetic acid, citric acid, and gluconic acid render some scale inhibitors, such as phosphonates, polycarboxylates, and phosphate esters, completely ineffective at concentrations as low as $0.1\ mg\,l^{-1}$. Such low concentrations may be expected to return from formation stimulation treatments for many months and would appear to jeopardize any scale inhibitor program in place.

This conclusion follows from experiments with a simulated North Sea scaling system at pH 4 and 6. The scale inhibitor concentrations studied were 50 and $100\ mg\,l^{-1}$. The large negative effect of the organic chelating agents was observed at pH 4 and 6. The only scale inhibitors studied that remained unaffected by these interferences were poly(vinyl sulfonate)s (PVS)s (Barthorpe, 1993).

MATHEMATICAL MODELS

Mathematical models have been developed (Shuler and Jenkins, 1989; Mackay et al., 1998; Mackay and Sorbie (1999, 1998)). The scale formation of iron carbonate and iron monosulfide has been simulated by thermodynamic and electrochemical models (Malandrino et al., 1998; Mackay and Sorbie, 1998; Anderko, 2000; Zhang et al., 2000). An accurate model to predict pH, scale indices, density, and inhibitor needs has been discussed. Experimental data to validate the model have been examined and an estimation of the error in analysis has been presented. Thus, a scale prediction software has been developed (Kan and Tomson, 2010).

The scaling tendency of sulfates, such as calcium sulfate, barite, and celestite, and further and halite scales are not a strong function of the pH of the brine. In contrast, carbonates, such as calcite, dolomite, and siderite, and sulfide scales are acid soluble. Therefore, their scaling tendencies are strongly dependent on the pH of the brine. For pH-sensitive scales, the scale prediction is more complicated (Kan and Tomson, 2010).

Optimal Dose

A method to estimate the optimal dose of a scale inhibitor has been described (Mikhailov et al., 1987). The method starts with noting the chemical composition and temperature of the water. From these parameters a stability index is calculated, allowing for the prediction of the optimal dose of a scale inhibitor.

The formation of calcium carbonate ($CaCO_3$), calcium sulfate ($CaSO_4$), and barium sulfate ($BaSO_4$) scales in brine may create problems with permeability. Therefore it is advantageous that newly made fractures have a scale inhibitor in place in the fracture to help prevent the formation of scale. Formulations of hydraulic fracturing fluids containing a scale inhibitor have been described in the literature (Watkins et al., 1993).

Precipitation Squeeze Method

In the precipitation squeeze method, the scale inhibitor reacts to form an insoluble salt which precipitates in the pores of the formation rock. For example, a phosphonate scale inhibitor and a calcium chelate are employed as a precipitation squeeze treatment. Further, phosphinic polycarboxylate has been used in a precipitation squeeze treatment. Poly(epoxysuccinic acid) is effective for the squeeze treatment (Brown and Brock, 1995).

An anionic scale inhibitor and a multivalent cation salt are dissolved in an alkaline aqueous liquid to provide a solution that contains both scale-inhibiting anions and multivalent cations which are mutually soluble under alkaline conditions. However, at lower pH the inhibitor is not soluble. One compound which reacts at a relatively slow rate to reduce the pH of the alkaline solution is dissolved in the solution. The rate at which the pH of the solution is reduced can be adjusted by the formulation (Collins, 2000).

Near-well squeeze treatment models assume that the flow pattern around the well is radial. It has been investigated whether strictly non-radial flow patterns around the well have a major effect on the squeeze treatment. It has been found that fractured wells have longer squeeze lifetimes than non-fractured wells.

Further, the calculations reveal that for fractured wells, inhibitor adsorption on the face of the fracture itself has no impact on the treatment lifetime. In a fractured well, the inhibitor is more retarded by contact with the rock over a greater distance in comparison to a matrix with radial treatment (Rakhimov et al., 2010).

INHIBITOR CHEMICALS

From the chemical classes, inhibitors can be coarsely subdivided in acids and complexing agents. Scale inhibitors described in the recent literature are summarized in Table 11.2.

TABLE 11.2 Scale Inhibitors

Compound	References
1-Hydroxyethylidene-1,1-diphosphonic acid	He et al. (1999)
Carbonic dihydrazide, $H_2N-NH-CO-NH-NH_2$	Mouche and Smyk (1995)
Polyaminealkylphosphonic acid and carboxymethyl cellulose or poly (acrylamide)	Kochnev et al. (1993)
Poly(acrylic acid) and chromium	Yan (1993)
Poly(acrylate)s[a]	Watkins et al. (1993)
Amine methylene phosphonate[b]	Graham et al. (2000)
Phosphonomethylated poly(amine)	Singleton et al. (2000)
Sulfonated poly(acrylate)copolymer	Chilcott et al. (2000)
Bis[tetrakis (hydroxymethyl) phosphonium] sulfate	Larsen et al. (2000)
Phosphonates	Holzner et al. (2000) and Jordan et al. (1997)
Carboxymethyl inulin	Kuzee and Raaijmakers (1999)
Polycarboxylic acid salts	Dobbs and Brown (1999)
Phosphoric acid esters of rice bran extract	Zeng and Fu (1998)
Poly(phosphino maleic anhydride)	Yang and Song (1998)
N,*N*-Diallyl-*N*-alkyl-*N*-(sulfoalkyl) ammonium betaine copolymer (with *N*-vinylpyrrolidone or acrylamide), diallylmethyltaurine hydrochloride ($CH_2{=}CH{-}CH_2Cl \times CH_3{-}NH{-}CH_2{-}CH_2{-}SO_3^-Na^+$)	Fong et al. (2001)
Aminotri(methylenephosphonic acid)	Kowalski and Pike (2001) and Tantayakom et al. (2004, 2005)
Diethylentrilopentrakis (methylenephosphonic acid)	Tantayakom et al. (2005)

[a]*In borate crosslinked fracturing fluids.*
[b]*High-temperature applications.*

Water-soluble Inhibitors

Acids

Both inorganic acids, such as hydrochloric acid and hydrofluoric acid, and organic acids, such as formic acid, can be used to increase the pH. Acids are used in combination with surfactants.

Acids, when used as scale inhibitors, are extremely corrosive. Their effectiveness has been tested in the laboratory. Parameters include acid type, metallurgy, temperature, inhibitor type and concentration, duration of acid-metal contact, and the effect of other chemical additives (Burger and Chesnut, 1992). Lead and zinc sulfide scale deposits can be removed by an acid treatment (Jordan et al., 2000).

Hydrofluoric Acid

It is known that permeability impairment may be improved by injecting acid formulations containing HF into the formation. Such methods are known to improve production from both subterranean calcareous and siliceous formations.

Most sandstone formations are composed of over 70% sand quartz, i.e., silica, bonded together by various amounts of cementing material including carbonate, dolomite, and silicates. Suitable silicates include clays and feldspars. A common method of treating sandstone formations involves introducing hydrofluoric acid into the wellbore and allowing the hydrofluoric acid to react with the surrounding formation.

Hydrofluoric acid exhibits a high reactivity toward siliceous minerals, such as clays and quartz fines. For instance, hydrofluoric acid reacts very quickly with authigenic clays, such as smectite, kaolinite, illite, and chlorite, especially at temperatures above 65 °C. As such, hydrofluoric acid is capable of attacking and dissolving siliceous minerals.

Upon contact of hydrofluoric acid with metallic ions present in the formation, such as sodium, potassium, calcium, and magnesium, undesirable precipitation reactions may occur.

Sandstone or siliceous formations and calcareous formations may be treated with an aqueous well treatment composition containing a hydrofluoric acid source in combination with a boron-containing compound and a phosphonate acid, ester, or salt. Such compositions have been shown to increase the permeability of the formation being treated by inhibiting or preventing the formation of undesirable inorganic scales, such as calcium fluoride, magnesium fluoride, potassium fluorosilicate, sodium fluorosilicate, or fluoroaluminate. As a result, the production from the formation is increased or improved (Ke and Qu, 2010).

Encapsulated Scale Inhibitors

This type of scale inhibitor allows chemical release over an extended period (Powell et al., 1995b; Hsu et al., 2000). Microencapsulated formulations may contain a gelatin coating with a multipurpose cocktail, such as Kowalski and Pike (1999, 2001):

- Scale inhibitor.
- Corrosion inhibitor.
- Biocide.
- Hydrogen sulfide scavengers.
- Demulsifier.
- Clay stabilizer.

TABLE 11.3 Chelating Agents for the Stabilization of Coatings (Kowalski and Pike, 1999)

Chelating agent	Acronym
N-(3-Hydroxypropyl) imino-*N*,*N*-diacetic acid	3-HPIDA
N-(2-Hydroxypropyl) imino-*N*,*N*-diacetic acid	2-HPIDA
N-Glycerylimino-*N*,*N*-diacetic acid	GLIDA
Dihydroxyisopropylimino-*N*,*N*-diacetic acid	DHPIDA
Methylimino-*N*,*N*-diacetic acid	MIDA
2-Methoxyethylimino-*N*,*N*-diacetic acid	MEIDA
Amidoiminodiacetic acid (= sodium amidonitrilo triacetic acid)	SAND
Acetamidoiminodiacetic acid	AIDA
3-Methoxypropylimino-*N*,*N*-diacetic acid	MEPIDA
Tris (hydroxymethyl) methylimino-*N*,*N*-diacetic acid	TRIDA

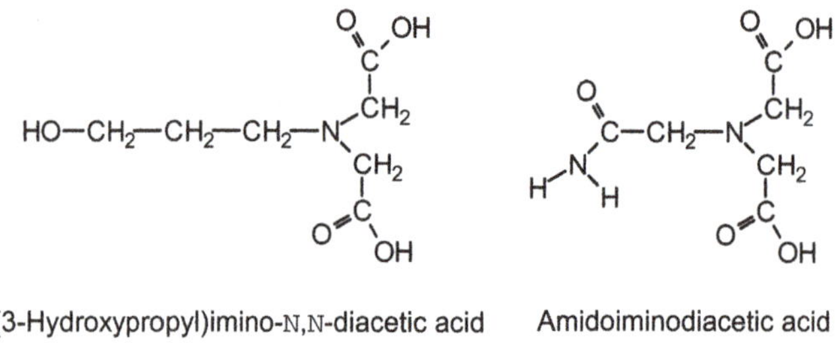

N-(3-Hydroxypropyl)imino-N,N-diacetic acid   Amidoiminodiacetic acid

FIGURE 11.1 Chelating agents.

Chelating Agents

Trace amounts of chelating agents, such as EDTA, citric acid, or gluconic acid, may lower the efficiency of scale inhibitors (Barthorpe, 1993). The concentration of calcium ions and magnesium ions affects the inhibition of barium sulfate (Boak et al., 1999). Pentaphosphonate, hexaphosphonate, phosphino-poly(carboxylic acid) (PPCA) salts, and PVS scale inhibitors have been studied. String chelating agents, given in Table 11.3, also stabilize the coating of encapsulated formulations (Kowalski and Pike, 1999). Some chelating agents based on imino acids are shown in Figure 11.1.

EDTA

A conventional scale dissolver for barite scale consists of a concentrated solution of potassium carbonate, potassium hydroxide, and the potassium salt of EDTA. On the other hand, carbonate scales may be dissolved using simple mineral acids, such as HCl (Jones et al., 2008). In addition, surfactants are advantageous for controlling the viscosity of the fluids. As surfactant, *N*-erucyl-*N,N*-bis (2-hydroxyethyl)-*N*-methyl ammonium chloride has been proposed (Jones et al., 2008).

These surfactants can form worm-like micelles when mixed with brines. The structure of the micelles contributes significantly to the viscoelasticity of the fluid, and the viscoelasticity is rapidly lost when the fluid contacts hydrocarbons, which cause the micelles to change structure or disband.

The difference in viscosity of the fluid when in contact with hydrocarbons and water allows a selective placement of the scale treatment. As a result, the scale may be preferentially removed from hydrocarbon-bearing zones. This can lead to a stimulation of hydrocarbon production without a substantial increase in the water cut of produced fluids (Jones et al., 2008).

By a suitable process, EDTA can be regenerated. Equation (11.1) illustrates the dissolution and subsequent isolation of a barium sulfate scale and the regeneration of EDTA in simplified form (Keatch, 2008).

$$\begin{aligned}
&\text{EDTA–K}_4 + \text{K}_2\text{CO}_3 + \text{BaSO}_4 \rightarrow \text{EDTA–K}_2\text{Ba} + \text{K}_2\text{CO}_3 + \text{K}_2\text{SO}_4,\\
&\text{K}_2\text{CO}_3 + 2\text{HCl} \rightarrow 2\text{KCl} + \text{H}_2\text{O} + \text{CO}_2,\\
&\text{EDTA–K}_2\text{Ba} + \text{K}_2\text{SO}_4 \rightarrow \text{EDTA–K}_4 + \text{BaSO}_4 \downarrow .
\end{aligned} \tag{11.1}$$

Phosphonates

Previous studies have strongly indicated that amine methylene phosphonic acid based inhibitor species, such as pentaphosphonate and hexaphosphonate, are considerably less thermally stable than polymeric species, such as PVS and the S-Co species. Therefore, the phosphonate-based species were reported as less applicable for deployment in high-temperature reservoir systems.

However, more recent inhibitor studies with species based on different amine methylene phosphonic acid revealed that certain species are thermally stable at temperatures exceeding 160 °C (Graham et al., 2002).

A series of phosphonate-based scale inhibitors were thermally aged at 160°C. After aging, the scale inhibitors were still able to prevent carbonate scale in dynamic tests. However, the performances of some of the phosphonate compounds against sulfate scale were reduced by thermal aging (Dyer et al., 2004).

Esterified phosphono or phosphino acids with a long-chain alcohol are effective as oil-soluble scale inhibitors, and as wax or asphaltene inhibitors or dispersants in oil production The esters can be prepared by the azeotropic condensation of the phosphino acids with the alcohol. Alternately, the esters are prepared by telomerizing an ester of an unsaturated carboxylic acid with a phosphite or hypophosphite telogen (Woodward et al., 2004).

In contrast, laboratory studies demonstrate a clear potential for a significant extension in treatment lifetime by changing from a phosphonate to a vinylsulfonate copolymer-based scale inhibitor (Jordan et al., 2005).

A solid, encapsulated scale inhibitor (calcium magnesium poly(phosphate)) has been developed and extensively tested for use in fracturing treatments (Powell et al., 1995a, 1995b, 1996). The inhibitor is compatible with borate crosslinked and zirconium crosslinked fracturing fluids and foamed fluids because of coating.

The coating exhibits a short-term effect on the release rate profile. The composition of the solid derivative has the greatest effect on its long-term release rate profile.

Alkaline Earth Sulfates

In dissolution studies of barite, using EDTA-based and diethylentriaminepentaacetic acid based chelating agents it has been verified that the presence of dicarboxylic acid additives, such as oxalate ion, improves the performance of the chelating agents. However, other related additives such as malonate and succinate reduce the effectiveness.

Oxalate ions catalyze the surface complexation reaction between the chelant and the barite surface by the formation of a two-ligand surface complex. The adverse effect observed at the other dicarboxylic acids is believed to arise due to steric effects, which prevent the formation of a two-ligand surface complex.

In extended studies with other barite-related scales, such as celestite ($SrSO_4$), gypsum ($CaSO_4 \times 2 H_2O$), and anhydrite ($CaSO_4$), it was observed that scale dissolvers which are optimized for their effectiveness against one type of scale, such as barite, may not be the most effective against other scales (Mendoza et al., 2002).

Biodegradable Scale Inhibitors

Many oil companies are requesting environmentally friendly fracturing fluids. Fracturing fluids are composed from a variety of compounds, each having a special function. Fracturing fluids contain also scale inhibitors. Here, we do not explain the basic issues of a fracturing fluid, this is explained in Chapter 1. Biodegradable chelants can be selected from a variety of compounds (Crews, 2006).

Sodium Iminodisuccinate

This compound is a maleic acid derivative. Its main use is as a chelant for divalent and trivalent ions. It complexes ions that can cause emulsions, form scale, can denature enzyme breakers, and cause crosslinked gel instability, and thus it can keep these ions from having these undesirable effects.

Disodium Hydroxyethyleneiminodiacetic Acid

This is one of the few amino carboxylic acid chelants that is readily biodegradable. It is useful for the chelation of divalent and trivalent ions that cause scale, can denature enzymes, and create crosslinked gel instability.

Sodium Gluconate and Sodium Glucoheptonate

These polyols are commonly used for chelation of mineral vitamins such as calcium, magnesium, iron, manganese, and copper. They have been also found to be useful herein to complex titanate, zirconate, and borate ions for crosslink delay purposes. They are also excellent iron complexors for enzyme breaker stability and crosslinked gel stability.

Sodium Poly(aspartate)

This compound is also known as polymerized aspartic amino acid. It chelates with multiple types of divalent and trivalent ions. It is useful in breaking emulsions and scale prevention.

Poly(aspartic acid)-based chemicals have been identified as environmentally friendly and biodegradable oilfield chemicals. They can be used both as corrosion inhibitors and scale inhibitors in brine-injection petroleum recovery. They exhibit a good calcium compatibility. At pH 5, poly(aspartate)s are resistant to calcium ion concentrations of 8500–7500 ppm, in comparison to a calcium ion concentration of 5000 ppm for phosphonate and maleic acid polymer products.

At a 5% concentration of the poly(aspartate)s, the calcium compatibility is superior to those of phosphonate and maleic acid polymer products. Poly(aspartate)s also do not interfere with the oil-water separation process (Fan et al., 2001).

Poly(aspartic acid) chemicals may not be exclusively used as scale inhibitor as such, but also in preconditioning solutions for other scale inhibitors. It has been claimed that a poly(aspartate) solution preconditioning solution at low pH advantageously enhances the adsorption of a phosphonate scale inhibitor to a rock material (Montgomerie et al., 2004).

It has been suggested to synthesize well treatment chemicals, such as poly(aspartate)s in a bioreactor at or near the site of the borehole. Still more straightforward, well treatment is achieved by introducing downhole thermophilic *Archea* or other thermophilic bacteria or organisms capable of generating well treatment chemicals (Kotlar and Haugan, 2005).

Oil-Soluble Scale Inhibitors

Basic compounds suitable for oil-soluble scale inhibitors include phosphonic acids, such as diethylenetriamine tetramethylene phosphonic acid, or bis-hexamethylene triamine pentakis (methylene phosphonic acid). Other suitable compounds are acrylic copolymers, PAA, PPCA, or phosphate esters. These basic compounds are blended with amine compounds to form an oil-soluble mix (Reizer et al., 2002). *tert*-Alkyl primary amines with 12–16 carbon atoms are oil soluble and effect the oil solubility of the scale inhibitor.

Aloe-based Scale Inhibitor

An aloe scale inhibitor composition is an aloe gel dissolved in water. The aloe gel comprises poly(saccharide)s, solubilized in water between 60 and 90 °C. In the chain carboxyl and alcohol functional groups are present that interact with divalent ions such as Ca^{2+} and Mg^{2+}.

Unlike chemically synthesized inhibitors the active ingredients in the aloe plant gel are naturally occurring compounds. The scale inhibitor can be applied at low and high calcium concentrations and without the limitation that the composition will precipitate because of hydrolysis. In contrast, hydrolysis favors the interaction with ions in the solution and, thus, its efficiency as a scale inhibitor may even increase (Viloria et al., 2010).

Reactivity toward calcium to form gels which encapsulate the calcium is believed to occur according to an egg-box model. The mechanism of trapping of the calcium ions is shown in Figure 11.2. In general, gels can be formed by the interaction of multivalent ions with polymers. This phenomenon is also known as physical crosslinking.

The chains of the gel interact with Ca^{2+} to get together. This causes stability when systemic forces or other conditions would otherwise try to revert the gel to an original condition.

The model assumes that calcium ions serve as a bridge to form ionic liaisons between two carboxyl groups belonging to two different chains in close contact. According to this poly(saccharide) model, the chains interact with Ca^{2+} allowing a structure coordinated packaging.

Acrolein Copolymer

The sulfide scale formation in petroleum prospecting and recovery can be reduced, by the injection of a copolymer of acrolein and ethylene during

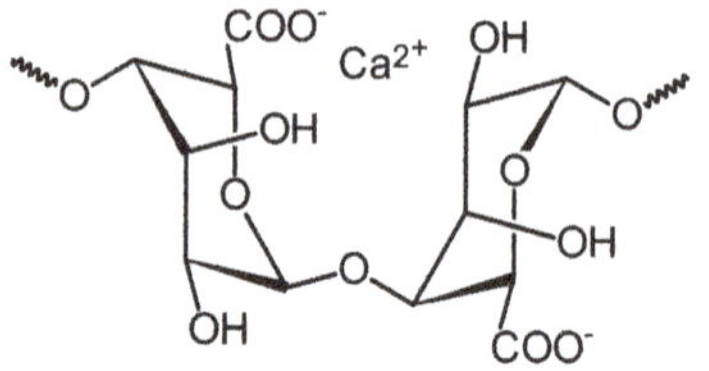

FIGURE 11.2 Egg-box model (Viloria et al., 2010).

hydraulic fracturing. In addition, an enhanced corrosion prevention occurs (Anon, 2010).

Following a baseline concentration of acrolein in the production fluid during recovery, the acrolein concentration is substantially steady after 24 h and gradually decreases over the following 6 months.

High Reservoir Temperatures

Conventional polymer and phosphonate scale inhibitors may not be appropriate for the application in high-pressure and high-temperature reservoirs. Only a limited range of commercially available oilfield scale inhibitor chemicals are sufficiently thermally stable at temperatures above 150 °C.

These chemicals are homopolymers of vinylsulfonate and copolymers of acrylic acid (AA) and vinylsulfonate. Other polymers, such as poly(maleic acid), poly(itaconic acid), and maleic acid/AA copolymers, may offer similar thermal stability (Collins, 1995). Thermal stability tests, influence on pH, ionic strength, and oxygen on conventional polymer and phosphonate scale inhibitors, for example, on phosphinopolycarboxylate, PVS, pentaphosphonate, and hexaphosphonate, have been presented (Graham et al., 1997, 1998b, 1998a; Dyer et al., 1999).

As pointed out above, it has been commonly believed that phosphonate scale inhibitors may not work for high-temperature inhibition applications, it has been more recently shown that phosphonate inhibitors are somehow effective at 200 °C under strictly anoxic conditions and in NaCl brines (Fan et al., 2010). In contrast, phosphonate inhibitors may precipitate with Ca^{2+} ions in a brine at high temperatures.

Tradenames in References

Tradename	Supplier
Description	
Dequest® 2060	Monsanto
Diethylene triamine pentamethylene phosphonic acid (Collins, 2000)	
Empol™ (Series)	Henkel
Oligomeric oleic acid (Jones et al., 2008)	
Rhodafac® RS-410	Rhodia
Poly(oxy-1,2-ethandiyl) tridecyl hydroxy phosphate (Martin etal., 2005)	
Scaletreat® XL14FD	TR Oil Services Ltd.
poly(maleate) (Collins, 2000)	

REFERENCES

Anderko, A., 2000. Simulation of $FeCO_3$/FeS scale formation using thermodynamic and electrochemical models. In: Proceedings Volume. NACE International Corrosion Conference, Corrosion 2000, Orlando, FL, 26–31 March 2000.

Anon, 2010. Process for reducing iron sulfide scales and preventing corrosion during the exploration for and production of hydrocarbons. IP.com J. 10 (12B), 22.

Barthorpe, R.T., 1993. The impairment of scale inhibitor function by commonly used organic anions. In: Proceedings Volume. SPE International Symposium on Oilfield Chemistry, New Orleans, 2–5 March 1993, pp. 69–76.

Becker, J.R., 1998. Corrosion and Scale Handbook. Pennwell Publishing Co, Tulsa.

Boak, L.S., Graham, G.M., Sorbie, K.S., 1999. The influence of divalent cations on the performance of $BaSO_4$ scale inhibitor species. In: Proceedings Volume. SPE International Symposium on Oilfield Chemistry, Houston, 16–19 February 1999, pp. 643–648.

Brown, J.M., Brock, G.F., 1995. Method of inhibiting reservoir scale. US Patent 5 409 062, assigned to Betz Laboratories, Inc., Trevose, PA, 25 April 1995. <http://www.freepatentsonline.com/5409062.html>.

Burger, E.D., Chesnut, G.R., 1992. Screening corrosion inhibitors used in acids for downhole scale removal. Mater. Perf. 31 (7), 40–44.

Chilcott, N.P., Phillips, D.A., Sanders, M.G., Collins, I.R., Gyani, A., 2000. The development and application of an accurate assay technique for sulphonated polyacrylate co- polymer oilfield scale inhibitors. In: Proceedings Volume. 2nd Annual SPE Oilfield Scale International Symposium, Aberdeen, Scotland, 26–27 January 2000.

Collins, I.R., 1995. Scale inhibition at high reservoir temperatures. In: Proceedings Volume. IBC Technical Services Ltd. Advances in Solving Oilfield Scaling International Conference, Aberdeen, Scotland, 20–21 November 1995.

Collins, I.R., 2000. Oil and gas field chemicals. US Patent 6 148 913, assigned to BP Chemicals Ltd., London, GB, 21 November 2000. <http://www.freepatentsonline.com/6148913.html>.

Crabtree, M., Eslinger, D., Fletcher, P., Miller, M., Johnson, A., King, G., 1999. Fighting scale—removal and prevention. Oilfield Rev. 11 (3), 30–45.

Crews, J.B., 2006. Biodegradable chelant compositions for fracturing fluid. US Patent 7 078 370, assigned to Baker Hughes Inc., Houston, TX, 18 July 2006. <http://www.freepatentsonline.com/7078370.html>.

Dobbs, J.B., Brown, J.M., 1999. An environmentally friendly scale inhibitor. In: Proceedings Volume. NACE International Corrosion Conference, Corrosion 99, San Antonio, 25–30 April 1999.

Duccini, Y., Dufour, A., Harm, W.M., Sanders, T.W., Weinstein, B., 1997. High performance oilfield scale inhibitors. In: Corrosion97. NACE International, New Orleans, LA. <http://www.onepetro.org/mslib/app/Preview.do?paperNumber=NACE-97169\&societyCode=NACE>.

Dyer, S.J., Graham, G.M., Sorbie, K.S., 1999. Factors affecting the thermal stability of conventional scale inhibitors for application in high pressure/high temperature reservoirs. In: Proceedings Volume. SPE International Symposium on Oilfield Chemistry, Houston, 16–19 February 1999, pp. 167–177.

Dyer, S.J., Anderson, C.E., Graham, G.M., 2004. Thermal stability of amine methyl phosphonate scale inhibitors. J. Petrol. Sci. Eng. 43, 259–270. http://dx.doi.org/10.1016/j.petrol.2004.02.018.

Fan, J.C., Fan, L.D.G., Liu, Q.W., Reyes, H., 2001. Thermal polyaspartates as dual function corrosion and mineral scale inhibitors. Polym. Mater. Sci. Eng. 84, 426–427. <http://www.onepetro.org/mslib/servlet/onepetropreview?id=00065005>.

Fan, C., Kan, A.T., Zhang, P., Lu, H., Work, S., Yu, J., Tomson, M.B., 2010. Scale prediction and inhibition for unconventional oil and gas production. In: SPE International Conference on Oilfield Scale. Society of Petroleum Engineers, Aberdeen, UK. <http://www.onepetro.org/mslib/app/Preview.do?paperNumber=SPE-130690-MS\&societyCode=SPE>.

Fong, D.W., Marth, C.F., Davis, R.V., 2001. Sulfobetaine-containing polymers and their utility as calcium carbonate scale inhibitors. US Patent 6 225 430, assigned to Nalco Chemical Co., 1 May 2001.

Frenier, W.W., Hill, D.G., 2004. Green inhibitors—development and applications for aqueous systems. In: Proceedings Volume. Volume 3 of Reviews on Corrosion Inhibitor Science and

Technology. Corrosion-2004 Symposium, New Orleans, LA, United States, 28 March–1 April, 2004, pp. 6/1–6/39.
Graham, G.M., Jordan, M.M., Sorbie, K.S., Bunney, J., Graham, G.C., Sablerolle, W., Hill, P., 1997. The implication of HP/HT (high pressure/high temperature) reservoir conditions on the selection and application of conventional scale inhibitors: Thermal stability studies. In: Proceedings Volume. SPE Oilfield International Symposium on Chemistry, Houston, 18–21 February 1997, pp. 627–640.
Graham, G.M., Dyer, S.J., Sorbie, K.S., Sablerolle, W., Graham, G.C., 1998a. Practical solutions to scaling in HP/HT (high pressure/high temperature) and high salinity reservoirs. In: Proceedings Volume. 4TH IBC UK Conf. Ltd Advances in Solving Oilfield Scaling International Conference, Aberdeen, Scotland, 28–29 January 1998.
Graham, G.M., Dyer, S.J., Sorbie, K.S., Sablerolle, W.R., Shone, P., Frigo, D., 1998b. Scale inhibitor selection for continuous and downhole squeeze application in HP/HT (high pressure/high temperature) conditions. In: Proceedings Volume. Annual SPE Technical Conference, New Orleans, 27–30 September 1998, pp. 645–659.
Graham, G.M., Dyer, S.J., Shone, P., 2000. Potential application of amine methylene phosphonate based inhibitor species in HP/HT (high pressure/high temperature) environments for improved carbonate scale inhibitor performance. In: Proceedings Volume. 2nd Annual SPE Oilfield Scale International Symposium, Aberdeen, Scotland, 26–27 January 2000.
Graham, G.M., Dyer, S.J., Shone, P., 2002. Potential application of amine methylene phosphonate-based inhibitor species in hp/ht environments for improved carbonate scale inhibitor performance. SPE Prod. Facil. 17, 212–220. <http://www.onepetro.org/mslib/servlet/onepetro preview?id=00060217>.
Hasson, D., Shemer, H., Sher, A., 2011. State of the art of friendly green scale control inhibitors: a review article. Ind. Eng. Chem. Res. 50 (12), 7601–7607. http://dx.doi.org/10.1021/ie200370v.
He, S., Kan, A.T., Tomson, M.B., 1999. Inhibition of calcium carbonate precipitation in NaCl brines from 25 to 90 °C. Appl. Geochem. 14 (1), 17–25
Holzner, C., Kleinstueck, R., Spaniol, A., 2000. Phosphonate-containing mixtures (Phosphonathaltige Mischungen). WO Patent 0032610, assigned to Bayer AG, 8 June 2000.
Hsu, J.F., Al-Zain, A.K., Raju, K.U., Henderson, A.P., 2000. Encapsulated scale inhibitor treatments experience in the ghawar field, saudi arabia. In: Proceedings Volume. 2nd Annual SPE Oilfield Scale International Symposium, Aberdeen, Scotland, 26–27 January 2000.
Jones, T.G.J., Tustin, G.J., Fletcher, P., Lee, J.C.-W., 2008. Scale dissolver fluid. US Patent 7343978, assigned to Schlumberger Technology Corp., Ridgefield, CT, 18 March 2008. <http://www.freepatentsonline.com/7343978.html>.
Jordan, M.M., Sorbie, K.S., Chen, P., Armitage, P., Hammond, P., Taylor, K., 1997. The design of polymer and phosphonate scale inhibitor precipitation treatments and the importance of precipitate solubility in extending squeeze lifetime. In: Proceedings Volume. SPE International Symposium on Oilfield Chemistry, Houston, 18–21 February 1997, pp. 641–651.
Jordan, M.M., Sjursaether, K., Bruce, R., Edgerton, M.C., 2000. Inhibition of lead and zinc sulphide scale deposits formed during production from high temperature oil and condensate reservoirs. In: Proceedings Volume. SPE Asia Pacific Oil & Gas Conference, Brisbane, Australia, 16–18 October 2000.
Jordan, M.M., Sjuraether, K., Collins, I.R., Feasey, N.D., Emmons, D., 2002. Life cycle management of scale control within subsea fields and its impact on flow assurance gulf of mexico and the North Sea basin. Spec. Publ. R. Soc. Lond. 280, 223–253. <http://www.onepetro.org/mslib/servlet/onepetropreview?id=00071557>.
Jordan, M.M., Kemp, S., Sorhaug, E., Sjursaether, K., Freer, B., 2003. Effective management of scaling from and within carbonate oil reservoirs, North Sea basin. Chem. Eng. Res. Des. 81, 359–372. http://dx.doi.org/10.1205/026387603605969l9.
Jordan, M.M., Sjursaether, K., Collins, I.R., 2005. Scale control within the North Sea chalk/limestone reservoirs-the challenge of understanding and optimizing chemical-placement methods and retention mechanisms: Laboratory to field. SPE Prod. Facil. 20, 262–273. <http://www.onepetro.org/mslib/servlet/onepetropreview?id=SPE-86476-PA>.
Kan, A.T., Tomson, M.B., 2010. Scale prediction for oil and gas production. In: International Oil and Gas Conference and Exhibition in China. Society of Petroleum Engineers, Beijing, China. <http://www.onepetro.org/mslib/app/Preview.do?paperNumber=SPE-132237-MS\&society Code=SPE>.

Ke, M., Qu, Q., 2010. Method for controlling inorganic fluoride scales. US Patent 7781381, assigned to BJ Services Company LLC, Houston, TX, 24 August 2010. <http://www.freepatentsonline.com/7781381.html>.

Keatch, R.W., 1998. Removal of sulphate scale from surface. GB Patent 2314865, 14 January 1998.

Keatch, R., 2008. Method for dissolving oilfield scale. US Patent 7470330, assigned to M-I Production Chemicals UK Ltd., Aberdeen, GB, Oilfield Mineral Solutions Ltd., Edinburgh, GB, 30 December 2008. <http://www.freepatentsonline.com/7470330.html>.

Kochnev, E.E., Merentsova, G.I., Andreeva, T.L., Ershov, V.A., 1993. Inhibitor solution to avoid inorganic salts deposition in oil drilling operations—contains water, carboxymethylcellulose or polyacrylamide and polyaminealkyl phosphonic acid and has improved distribution uniformity. SU Patent 1787996, assigned to Siberian Research Institute of the Oil Industry, 15 January 1993.

Kotlar, H.K., Haugan, J.A., 2005. Genetically engineered well treatment microorganisms. GB Patent 2413797, assigned to Statoil Asa, 9 November 2005.

Kowalski, T.C., Pike, R.W., 1999. Microencapsulated oil field chemicals. US Patent 5922652, 13 July 1999.

Kowalski, T.C., Pike, R.W., 2001. Microencapsulated oil field chemicals. US Patent 6326335, assigned to Corsicana Technologies Inc., 4 December 2001.

Kuzee, H.C., Raaijmakers, H.W.C., 1999. Method for preventing deposits in oil extraction. WO Patent 9964716, assigned to Cooperatie Cosun Ua, 16 December 1999.

Larsen, J., Sanders, P.F., Talbot, R.E., 2000. Experience with the use of tetrakishydroxymethylphosphonium sulfate (THPS) for the control of downhole hydrogen sulfide. In: Proceedings Volume. NACE International Corrosion Conference, Corrosion 2000, Orlando, FL, 26–31 March 2000.

Mackay, E.J., Sorbie, K.S., 1998. Modelling scale inhibitor squeeze treatments in high crossflow horizontal wells. J. Can. Pet. Technol. 39 (10), 47–51.

Mackay, E.J., Sorbie, K.S., 1999. An evaluation of simulation techniques for modelling squeeze treatments. In: Proceedings Volume. Annual SPE Technical Conference, Houston, 3–6 October 1999, pp. 373–387.

Mackay, E.J., Sorbie, K.S., Jordan, M.M., Matharu, A.P., Tomlins, R., 1998. Modelling of scale inhibitor treatments in horizontal wells: Application to the alba field. In: Proceedings Volume. SPE International Symposium on Formation Damage Control, Lafayette, LA, 18–19 February 1998, pp. 337–348.

Malandrino, A., Andrei, M., Gagliardi, F., Lockhart, T.P., 1998. A thermodynamic model for PPCA (phosphino-polycarboxylic acid) precipitation. In: Proceedings Volume. 4th IBC UK Conf. Ltd Advances in Solving Oilfield Scaling International Conference, Aberdeen, Scotland, 28–29 January 1998.

Martin, R.L., Brock, G.F., Dobbs, J.B., 2005. Corrosion inhibitors and methods of use. US Patent 6866797, assigned to BJ Services Co., 15 March 2005. <http://www.freepatentsonline.com/6866797.html>.

Mendoza, A., Graham, G.M., Farquhar, M.L., Sorbie, K.S., 2002. Controlling factors of EDTA and DTPA based scale dissolvers against sulphate scale. Prog. Min. Oilfield Chem. 4, 41–58.

Mikhailov, S.A., Khmeleva, E.P., Moiseeva, E.V., Sleta, T.M., 1987. Determination of the optimal dose of salt deposition inhibitors. Neft Khoz 7, 43–45.

Montgomerie, H.T.R., Chen, P., Hagen, T., Wat, R.M.S., Selle, O.M., Kotlar, H.K., 2004. Method of controlling scale formation. WO Patent 2004011772, assigned to Champion Technology Inc., Statoil Asa, Montgomerie Harry Trenouth Rus, Chen Ping, Hagen Thomas, Wat Rex Man Shing, Selle Olav Martin, and Kotlar Hans Kristian, 5 February 2004.

Mouche, R.J., Smyk, E.B., 1995. Noncorrosive scale inhibitor additive in geothermal wells. US Patent 5403493, assigned to Nalco Chemical Co., 4 April 1995.

Powell, P.J., Gdanski, R.D., McCabe, M.A., Buster, D.C., 1995a. Controlled-release scale inhibitor for use in fracturing treatments. In: Proceedings Volume. SPE Oilfield International Symposium on Chemistry, San Antonio, 14–17 February 95, pp. 571–579.

Powell, R.J., Fischer, A.R., Gdanski, R.D., McCabe, M.A., Pelley, S.D., 1995b. Encapsulated scale inhibitor for use in fracturing treatments. In: Proceedings Volume. Annual SPE Technical Conference, Dallas, 22–25 October 1995, pp. 557–563.

Powell, R.J., Fischer, A.R., Gdanski, R.D., McCabe, M.A., Pelley, S.D., 1996. Encapsulated scale inhibitor for use in fracturing treatments. In: Proceedings Volume. SPE Permian Basin Oil & Gas Recovery Conference, Midland, TX, 27–29 March 1996, pp. 107–113.

Rakhimov, A.Z., Vazquez, O., Sorbie, K.S., Mackay, E.J., 2010. Impact of fluid distribution on scale inhibitor squeeze treatments. In: SPE EUROPEC/EAGE Annual Conference and Exhibition. Society of Petroleum Engineers, Barcelona, Spain. <http://www.onepetro.org/mslib/app/Preview.do?paperNumber=SPE-131724-MS\&societyCode=SPE>.

Reizer, J.M., Rudel, M.G., Sitz, C.D., Wat, R.M.S., Montgomerie, H., 2002. Scale inhibitors. US Patent 6 379 612, assigned to Champion Technology Inc., 30 April 2002.

Shuler, P.J., Jenkins, W.H., 1989. Prevention of downhole scale deposition in the ninian field. In: Proceedings Volume. Vol. 2. SPE Offshore Europe Conference, Aberdeen, Scotland, 5–8 September 1989.

Sikes, C.S., Wierzbicki, A., 1996. Stereospecific and nonspecific inhibition of mineral scale and ice formation. In: Proceedings Volume. 51st Annual NACE International Corrosion Conference, Corrosion 96, Denver, 24–29 March 1996.

Singleton, M.A., Collins, J.A., Poynton, N., Formston, H.J., 2000. Developments in phosphonomethylated polyamine (PMPA) scale inhibitor chemistry for severe $BaSO_4$ scaling conditions. In: Proceedings Volume. 2nd Annual SPE Oilfield Scale International Symposium, Aberdeen, Scotland, 26–27 January 2000.

Talbot, R.E., Jones, C.R., Hills, E., 2009. Scale inhibition in water systems. US Patent 7 572 381, assigned to Rhodia UK Ltd., Hertfordshire, GB, 11 August 2009. <http://www.freepatentsonline.com/7572381.html>.

Tantayakom, V., Fogler, H.S., de Moraes, F.F., Bualuang, M., Chavadej, S., Malakul, P., 2004. Study of Ca-ATMP precipitation in the presence of magnesium ion. Langmuir 20, 2220–2226. <http://pubs.acs.org/doi/abs/10.1021/la0358318>.

Tantayakom, V., Fogler, H.S., Chavadej, S., 2005. Scale inhibitor precipitation kinetics. In: Proceedings Volume. 7th World Congress of Chemical Engineering, Glasgow, United Kingdom, 10–14 July 2005, pp. 85704/1–85704/8.

Viloria, A., Castillo, L., Garcia, J.A., Biomorgi, J., 2010. Aloe derived scale inhibitor. US Patent 7 645 722, assigned to Intevep, S.A., Caracas, VE, 12 January 2010. <http://www.freepatentsonline.com/7645722.html>.

Watkins, D.R., Clemens, J.J., Smith, J.C., Sharma, S.N., Edwards, H.G., 1993. Use of scale inhibitors in hydraulic fracture fluids to prevent scale build-up. US Patent 5 224 543, assigned to Union Oil Co., California, 6 July 1993.

Woodward, G., Jones, C.R., Davis, K.P., 2004. Novel phosphonocarboxylic acid esters. WO Patent 2 004 002 994, assigned to Rhodia Consumer Specialities L, Woodward Gary, Jones Christopher Raymond, and Davis Keith Philip, 8 January 2004.

Yan, T.Y., 1993. Process for inhibiting scale formation in subterranean formations. WO Patent 9 305 270, assigned to Mobil Oil Corp., 18 March 1993.

Yang, L., Song, B., 1998. Phosphino maleic anhydride polymer as scale inhibitor for oil/gas field produced waters. Oilfield Chem. 15 (2), 137–140

Zeng, Y.B., Fu, S.B., 1998. The inhibiting property of phosphoric acid esters of rice bran extract for barium sulfate scaling. Oilfield Chem. 15 (4), 333–335, 365.

Zhang, H., Mackay, E.J., Sorbie, K.S., Chen, P., 2000. Non-equilibrium adsorption and precipitation of scale inhibitors: corefloods and mathematical modelling. In: Proceedings Volume. SPE International Oil & Gas Conference and Exhibition in China, Beijing, 7–10 November 2000, pp. 1–18.

Chapter 12

Foaming Agents

Foamed fluids can be used in many fracturing jobs, especially when environmental sensitivity is a concern (Stacy and Weber, 1995). Foam-fluid formulations are reusable, shear stable, and form stable foams over a wide temperature range. They exhibit high viscosities even at relatively high temperatures (Bonekamp et al., 1993).

Foamed fracturing fluids are superior to conventional liquid fracturing fluids for problematic and water sensitive formations because foams contain substantially less liquid than liquid fracturing fluids and therefore have less tendency to leak. Also, the use of foams requires less liquid retrieval after the fracturing operation is complete. Moreover, the sudden expansion of the gas in the foams from pressure in the well being relieved after the fracturing operation is complete promotes the flow of the residual fracture fluid liquid back into the well.

The foamed fracturing fluid may also include a proppant material for preventing formed fractures from closing. A variety of proppant materials can be utilized including resin-coated or uncoated sand, sintered bauxite, ceramic materials, and glass beads. The proppant material is preferably used in an amount in the range from about 1 lb to 10 lb of proppant material per gallon of the foamed fracturing fluid (Dahanayake et al., 2008).

The content of the gas is called *quality*. The foam quality Q is expressed as a percentage as shown in Eq. (12.1).

$$Q = 100\frac{V_f - V_l}{V_f}. \tag{12.1}$$

Here, V_f is the total volume of the foam and V_l is the volume of the liquid in the foam. Therefore, a 70 quality contains 70% gas. Recently, foams with 95% gas have been examined. For such foam types, only a foam prepared from 2% of an anionic surfactant with plain water had a uniform, fine-bubble structure (Harris and Heath, 1996).

Surfactants are available that change the power of foaming. For example, a tertiary alkyl amine ethoxylate can change in the character from a foaming surfactant to a nonfoaming surfactant by lowering the pH of the environment,

Hydraulic Fracturing Chemicals and Fluids Technology. http://dx.doi.org/10.1016/B978-0-12-411491-3.00012-1
© 2013 Elsevier Inc. All rights reserved.

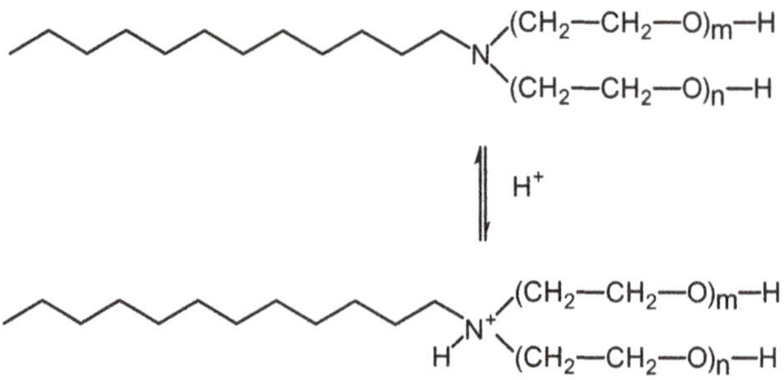

FIGURE 12.1 Changing the foaming ability by changing the pH. Top: Foaming modification. Bottom: Non-foaming modification (Welton et al., 2010).

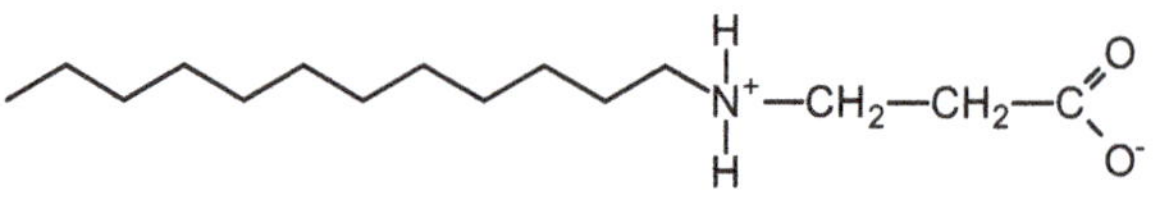

FIGURE 12.2 Lauryl betaine.

by adding hydrogen ions. It can then be changed back to a foaming surfactant by the addition of a basic material, e.g., hydroxide ions. At low pH the amine group is quaternized, as shown in Figure 12.1.

In addition, cocobetaine and α-olefin sulfonate have been proposed as foamers (Pakulski and Hlidek, 1992). A preferred amphoteric surfactant is a mixture of laurylamine and myristylamine oxide. Lauryl betaine is shown in Figure 12.2. Foamed fracturing fluids can be recycled (Chatterji et al., 2007) as the pH of the fracturing fluid is changed so that the foam is destroyed.

ENVIRONMENTALLY SAFE FLUIDS

Conventional foamed fracturing fluids have included various surfactants known as foaming and foam stabilizing agents for facilitating the foaming and stabilization of the foam produced when a gas is mixed with the fracturing fluid (Chatterji et al., 2004).

However, the foaming and stabilizing surfactants have not met complete environmental requirements. That is, when the foaming and stabilizing surfactants find their way into water in the environment, they do not fully degrade which can result in interference with aquatic life cycles.

An environmentally harmless hydrolyzed keratin additive for foaming and stabilizing a gelled water fracturing fluid can be manufactured by the hydrolysis of hoof and horn meal. There, the hoof and horn meal is heated with lime in an autoclave to produce a hydrolyzed protein. Such a protein is commercially available as a free-flowing powder that contains about 85% protein.

The non-protein portion of the powder consists of about 0.58% insoluble material with the remainder being soluble non-protein materials primarily made up of calcium sulfate, magnesium sulfate, and potassium sulfate (Chatterji et al., 2004).

When the viscosity and stability of a foamed fracturing fluid should be even further increased, a foam viscosity and stability increasing additive can be included in the foamed fracturing fluid. Examples of such foam viscosity and stability increasing additives include iodine, hydrogen peroxide, cupric sulfate, and zinc bromide.

LIQUID CARBON DIOXIDE FOAMS

Foamed nitrogen in liquid CO_2 can be used for fracturing purposes (Gupta et al., 2004). The foam forming substance is preferably a nonionic hydrofluoroether surfactant. The liquid phase is CO_2 and the gaseous phase is N_2.

Perfluorinated compounds can be synthesized by direct fluorination, electrochemical fluorination, addition polymerization of fluorine-containing monomers, and oxidative polymerization of fluorine-containing monomers.

Perfluorinated compounds, since they lack chlorine atoms, are not ozone-depleting agents, but these compounds may exhibit a global warming potential due to their long atmospheric lifetimes. Therefore, it is preferred that the fluorine-based stabilizer contains at least one aliphatic hydrogen atom in the molecule. These compounds generally are very thermally and chemically stable, yet are much more environmentally acceptable in that they degrade in the atmosphere and thus have a low global warming potential, in addition to a zero ozone depletion potential. An example of such a compound is hexafluoroxylene.

Fluorinated ethers can be prepared by alkylation of perfluorinated alkoxides. These are prepared by the reaction of the corresponding perfluorinated acyl fluoride or perfluorinated ketone with an anhydrous alkali metal fluoride or anhydrous silver fluoride in an anhydrous polar, aprotic solvent. Alternatively, a fluorinated tertiary alcohol can be allowed to react with a base, e.g., potassium hydroxide or sodium hydride, to produce a perfluorinated tertiary alkoxide which can then be alkylated by the reaction with an alkylating agent. Examples of hydrofluoroethers are methoxy nonafluorobutane, or ethoxy nonafluorobutane.

The methods of preparation of formulations have been described in detail. It has been found that fairly high-viscosity foams, with high proppant loading characteristics can be formed (Gupta et al., 2004).

Tradenames in References

Tradename	Supplier
Description	
Flow-Back™	Halliburton Energy Services, Inc.
Xanthan gum, welan gum (Chatterji et al., 2004)	
Fluorinert™ (Series)	3M Comp.
Fluorocarbons (Gupta et al., 2004)	
Flutec™ PP	BNFL Fluorochemicals Ltd.
Fluorocarbon (Gupta et al., 2004)	
Galden™ LS	Montedison Inc.
Fluorocarbon (Gupta et al., 2004)	
Krytox™	DuPont
Fluorinated oils and greases (Gupta et al., 2004)	
WS-44	Halliburton Energy Services, Inc.
Emulsifier (Welton et al., 2010)	

REFERENCES

Bonekamp, J.E., Rose, G.D., Schmidt, D.L., Teot, A.S., Watkins, E.K., 1993. Viscoelastic surfactant based foam fluids. US Patent 5 258 137, assigned to Dow Chemical Co., 2 November 1993.

Chatterji, J., Crook, R., King, K.L., 2004. Foamed fracturing fluids, additives and methods of fracturing subterranean zones. US Patent 6 734 146, assigned to Halliburton Energy Services, Inc., Duncan, OK, 11 May 2004. <http://www.freepatentsonline.com/6734146.html>.

Chatterji, J., King, B.J., King, K.L., 2007. Recyclable foamed fracturing fluids and methods of using the same. US Patent 7 205 263, assigned to Halliburton energy Services, Inc., Duncan, OK, 17 April 2007. <http://www.freepatentsonline.com/7205263.html>.
xmllabelb0020

Dahanayake, M.S., Kesavan, S., Colaco, A., 2008. Method of recycling fracturing fluids using a self-degrading foaming composition. US Patent 7 404 442, assigned to Rhodia Inc., Cranbury, NJ, 29 July 2008. <http://www.freepatentsonline.com/7404442.html>.

Gupta, D.V.S., Pierce, R.G., Senger Elsbernd, C.L., 2004. Foamed nitrogen in liquid CO_2 for fracturing. US Patent 6 729 409, 4 May 2004. <http://www.freepatentsonline.com/6729409.html>.

Harris, P.C., Heath, S.J., 1996. High-quality foam fracturing fluids. In: Proceedings Volume. SPE Gas Technology Symposium, Calgary, Canada 28 April–1 May 1996, pp. 265–273.

Pakulski, M.K., Hlidek, B.T., 1992. Slurried polymer foam system and method for the use thereof. WO Patent 9 214 907, assigned to Western Co., North America, 3 September 1992.

Stacy, A.L., Weber, R.B., 1995. Method for reducing deleterious environmental impact of subterranean fracturng processes. US Patent 5 424 285, assigned to Western Co., North America, 13 June 1995.

Welton, T.D., Todd, B.L., McMechan, D., 2010. Methods for effecting controlled break in pH dependent foamed fracturing fluid. US Patent 7 662 756, assigned to Halliburton Energy Services, Inc., Duncan, OK, 16 February 2010. <http://www.freepatentsonline.com/7662756.html>.

Chapter 13

Defoamers

Defoaming is necessary in several industrial branches and is often a key factor for efficient operation. A review on defoamers has been given by Owen (1996).

THEORY OF DEFOAMING

Stability of Foams

Foams are thermodynamically unstable but are prevented from collapsing by the following properties:

- Surface elasticity,
- Viscous drainage,
- Reduced gas diffusion between bubbles, and
- Thin-film stabilization effects from the interaction of opposite surfaces.

The stability of a foam can be explained by the Gibbs elasticity E. The Gibbs elasticity results from reducing the surface concentration of the active molecules in equilibrium when the film is extended. This causes an increase in the equilibrium surface tension σ, which acts as a restoring force

$$E = 2A\frac{d\sigma}{dA}. \tag{13.1}$$

A is the area of the surface. In a foam, where the surfaces are interconnected, the time-dependent Marangoni effect is important. A restoring force corresponding to the Gibbs elasticity will appear, because only a finite rate of absorption of the surface active agent, which decreases the surface tension, can take place on the expansion and contraction of a foam. Thus the Marangoni effect is a kinetic effect.

The surface tension effects under nonequilibrium conditions are described in terms of dilatational moduli. The complex dilatational modulus ε of a single surface is defined in the same way as the Gibbs elasticity. The factor 2 is not used in a single surface

$$\varepsilon = 2A\frac{d\sigma}{dA}. \tag{13.2}$$

Hydraulic Fracturing Chemicals and Fluids Technology. http://dx.doi.org/10.1016/B978-0-12-411491-3.00013-3
© 2013 Elsevier Inc. All rights reserved.

In a periodic dilatational experiment, the complex elasticity module is a function of the angular frequency:

$$\varepsilon\left(i\omega\right) = |\varepsilon| \cos\theta + i\,|\varepsilon| \sin\theta = \varepsilon_d\left(\omega\right) + \omega\eta_d\left(\omega\right). \tag{13.3}$$

ε_d is the dilatational elasticity, and η_d is the dilatational viscosity. It is characteristic for a stable foam to exhibit a high surface dilatational elasticity and a high dilatational viscosity. Therefore, effective defoamers should reduce these properties of the foam.

Under nonequilibrium conditions, both a high bulk viscosity and a surface viscosity can delay the film thinning and the stretching deformation, which precedes the destruction of a foam. There is another issue that concerns the formation of ordered structures. The development of ordered structures in the surface film may also stabilize the foams. Liquid crystalline phases in surfaces enhance the stability of the foam.

If the gas diffusion between bubbles is reduced, the collapse of the bubbles is delayed by retarding the bubble size changes and the resulting mechanical stresses. Therefore, single films can persist longer than the corresponding foams. However, this effect is of minor importance in practical situations. Electric effects, such as those occuring in double layers, form opposite surfaces of importance only for extremely thin films, less than 10 nm. In particular, they occur with ionic surfactants.

Action of Defoamers

At high bulk viscosity, lowering the surface tension is not relevant for the mechanism of stabilization of foams, but for all other mechanisms of foam stabilization a change of the surface properties is essential. A defoaming agent will change the surface properties of a foam upon activation. Most defoamers have a surface tension in the range of 20–30 m Nm^{-1}. The surface tensions of some defoamers are shown in Table 13.1.

TABLE 13.1 Surface Tensions of Some Defoamers

Material	Surface Tension at 20 °C (m Nm^{-1})
Poly(oxypropylene) 3 kDa	31.2
Poly(dimethylsiloxane) 3.9 kDa	20.2
Mineral oil	28.8
Corn oil	33.4
Peanut oil	35.5
Tributyl phosphate	25.1

Two related antifoam mechanisms have been proposed for low surface tensions of certain defoamer formulations:

1. The defoamer is dispersed in fine droplets in the liquid. From the droplets, the molecules may enter the surface of the foam. The tensions created by this spreading result in the eventual rupture of the film.
2. Alternatively, it is suggested that the molecules will form a monolayer rather than spreading. The monolayer has less coherence than the original monolayer on the film and causes a destabilization of the film.

Spreading Coefficient

The spreading coefficient is defined as the difference of the surface tension of the foaming medium σ_f, the surface tension of the defoamer σ_d, and the interfacial tension of both materials σ_{df}

$$S = \sigma_f - \sigma_d - \sigma_{df}. \tag{13.4}$$

It can be readily seen that the spreading coefficient S becomes increasingly positive as the surface tension of the defoamer becomes smaller. This indicates the thermodynamic tendency of defoaming.

The above statements are adequate for liquid defoamers that are insoluble in the bulk. However, experience has proven that certain dispersed hydrophobic solids can greatly enhance the effectiveness of defoaming. A strong correlation between the effectiveness of a defoamer and the contact angle for silicone-treated silica in hydrocarbons has been established. It is believed that the dewetting process of the hydrophobic silica causes the collapse of a foam by the direct mechanical shock occurring by this process.

CLASSIFICATION OF DEFOAMERS

Defoamer formulations contain numerous ingredients to meet the diverse requirements for which they are formulated. Various classification approaches are possible, including the classification by application, physical form of the defoamer, and the chemical type of the defoamer. In general, defoamers contain a variety of active ingredients, both in solid and in liquid states, and a number of ancillary agents such as emulsifiers, spreading agents, thickeners, preservatives, carrier oils, compatibilizers, solvents, and water.

Active Ingredients

Active ingredients are the components of the formulation that control the actual foaming. These may be liquids or solids.

Liquid Components

Because lowering the surface tension is the most important physical property of a defoamer, it is reasonable to classify the defoamer by the hydrophobic operation of the molecule. In contrast, the classification of organic molecules by functional groups is often polar and hydrophilic, i.e., alcohol, acid, and salt are common in basic organic chemistry. Four classes of defoamers are known as liquid phase components:

- Hydrocarbons,
- Poly(ether)s,
- Silicones, and
- Fluorocarbons.

Synergistic Antifoam Action by Solid Particles

Often, dispersed solids are active in defoaming in suitable formulations. Some liquid defoamers are believed to be active only in the presence of a solid. It is believed that a surface active agent present in the system will carry the solid particles in the region of the interface and the solid will cause a destabilization of the foam.

For example, a synergistic defoaming occurs when hydrophobic solid particles are used in conjunction with a liquid that is insoluble in the foamy solution (Frye and Berg, 1989). Mechanisms for film rupture by either the solid or the liquid alone have been elucidated, along with explanations for the poor effectiveness, which are observed with many foam systems for these single-component defoamers.

Silicone Antifoaming Agents

Poly(dimethylsiloxane) is active in nonaqueous systems, but it shows little foam-inhibiting effect in aqueous systems. However, when it is compounded with a hydrophobic-modified silica, a highly active defoamer emerges.

Several factors contribute to the dual nature of silicone defoamers. For example, soluble silicones can concentrate at the air-oil interface to stabilize bubbles, while dispersed drops of silicone can accelerate the coalescence process by rapidly spreading at the gas-liquid interface of a bubble, causing film thinning by surface transport (Mannheimer, 1992).

Silicones exhibit an apparently low solubility in different oils. In fact, there is actually a slow rate of dissolution that depends on the viscosity of the oil and the concentration of the dispersed drops. The mechanisms of the critical bubble size and the reason for a significantly faster coalescence at a lower concentration of silicone can be explained in terms of the higher interfacial mobility, as can be measured by the bubble rise velocities.

TABLE 13.2 Composition of a Defoamer for Hydraulic Fracturing Fluids

Compound	Amount (%)
C_6-C_{12} mixture of polar compounds	50–90
Sorbitan monooleate	10–50
Polyglycol M = 3.8 kDa	10

Exemplary Composition

A defoamer and an antifoamer composition are described for defoaming aqueous fluid systems (Zychal, 1986). The composition of a typical defoamer for hydraulic fracturing fluids is shown in Table 13.2.

Orthoesters are added that will generate acids in order to degrade the foam. Examples of suitable orthoesters and poly(orthoesters) are trimethyl orthoacetate, triethyl orthoacetate, and the corresponding orthoformates. By the way, poly(orthoesters) are important in medical applications (Heller et al., 2002). Some simple orthoesters are shown in Figure 13.1.

The synthesis of orthoesters may proceed either by a Williamson synthesis or by the addition of alcohols to a cyanide. The respective reactions are shown in Figure 13.2.

Orthoesters are stable toward alkalis, but not stable toward acids and water. The orthoester decreases the pH of the foamed fracturing fluid to sufficiently convert the foaming surfactant to a nonfoaming surfactant, whereby the foamed fracturing fluid substantially defoams. To allow the orthoester to hydrolyze to produce an acid, a source of water is needed, whether from the formation or introduced into the formation. The water should be present in an amount of 2 mol of water per mol of orthoester.

When the orthoester ultimately hydrolyzes and generates the acid, the acid may react with the foaming surfactant making it to become predominantly a nonfoaming surfactant (Welton et al., 2010). The orthoester compositions may also contain an inhibitor, which may delay the generation of the acid from the

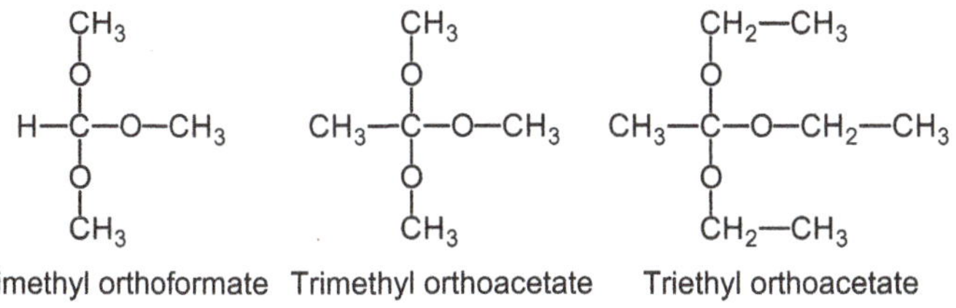

FIGURE 13.1 Orthoesters.

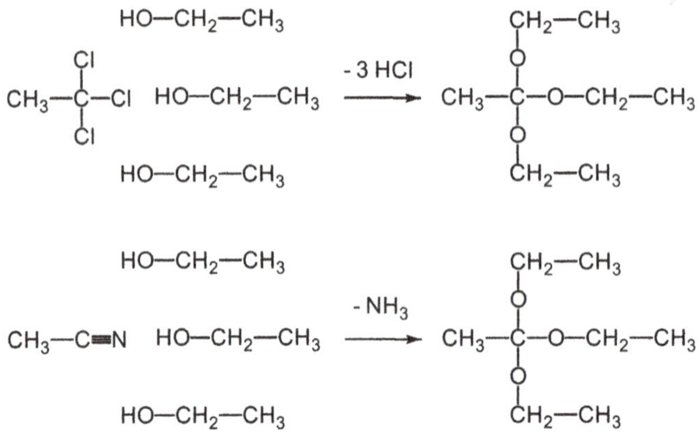

$$CH_3{-}CCl_3 \quad 3\ HO{-}CH_2{-}CH_3 \xrightarrow{-3\,HCl} CH_3{-}C(O{-}CH_2{-}CH_3)_3$$

$$CH_3{-}C{\equiv}N \quad 3\ HO{-}CH_2{-}CH_3 \xrightarrow{-NH_3} CH_3{-}C(O{-}CH_2{-}CH_3)_3$$

FIGURE 13.2 Synthesis of orthoesters.

orthoester of the orthoester composition. In addition, they may neutralize any generated acid during the delay period. Suitable inhibitors include bases., e.g., alkali hydroxides, sodium carbonate, or hexamethylenetetramine.

Sometimes, a small amount of a strong base as opposed to a large amount of a relatively weak base is preferred to achieve the delayed generation of the acid and the neutralization of the generated acid for a desired delay period. A foamed fracturing composition may in addition contain other usual ingredients, such as (Welton et al., 2010):

- Gelling agents,
- Bactericides, and
- Proppants.

Tradenames in References

Tradename	Supplier
Description	
WS-44	Halliburton Energy Services, Inc.
Emulsifier (Welton et al., 2010)	

REFERENCES

Frye, G.C., Berg, J.C., 1989. Mechanisms for the synergistic antifoam action by hydrophobic solid particles in insoluble liquids. J. Colloid Interface Sci. 130 (1), 54–59.

Heller, J., Barr, J., Ng, S.Y., Abdellauoi, K.S., Gurny, R., 2002. Poly(ortho esters): synthesis, characterization, properties and uses. Adv. Drug Deliv. Rev. 54 (7), 1015–1039. <http://www.sciencedirect.com/science/article/B6T3R-46XH%20K44-4/2/fec170fd72f87dc13b7290e749af3388>.

Mannheimer, R.J., 1992. Factors that influence the coalescence of bubbles in oils that contain silicone antifoamants. Chem. Eng. Commun. 113, 183–196.

Owen, M.J., 1996. Defoamers, fourth ed. In: Kirk-Othmer (Ed.), Encyclopedia of Chemical Technology, vol. 7. John Wiley and Sons, New York, Chichester, Brisbane, pp. 929–945.

Welton, T.D., Todd, B.L., McMechan, D., 2010. Methods for effecting controlled break in pH dependent foamed fracturing fluid. US Patent 7662756, assigned to Halliburton Energy Services, Inc., Duncan, OK, 16 February 2010. <http://www.freepatentsonline.com/7662756.html>.

Zychal, C., 1986. Defoamer and antifoamer composition and method for defoaming aqueous fluid systems. US Patent 4631145, assigned to Amoco Corp., 23 December 1986.

Chapter 14

Crosslinking Agents

KINETICS OF CROSSLINKING

The rheology of hydroxypropyl guar is greatly complicated by the crosslinking reactions with titanium ions. A study to better understand the rheology of the reaction of hydroxypropyl guar with titanium chelates and how the rheology depends on the residence time, shear history, and chemical composition has been performed (Barkat, 1987).

Rheologic experiments were performed to obtain information about the kinetics of crosslinking in hydroxypropyl guar. Continuous flow and dynamic data suggest a crosslinking reaction order of approximately 4/3 and 2/3, respectively, with respect to the crosslinker and hydroxypropyl guar concentration. Dynamic tests have shown that the shearing time is important in determining the final gel properties.

Continued steady shear and dynamic tests show that high shear irreversibly destroys the gel structure, and the extent of the crosslinking reaction decreases with increasing shear. Studies at shear rates below $100\,s^{-1}$ suggest a shear-induced structural change in the polymer that affects the chemistry of the reaction and the nature of the product molecule.

Delayed Crosslinking

Delayed crosslinking is desirable because the fluid can be pumped down more easily. A delay is a retarded reaction rate of crosslinking. This can be achieved with the methods explained subsequently.

CROSSLINKING ADDITIVES

Borate Systems

Boric acid can form complexes with hydroxyl compounds. The mechanism of formation complexes of boric acid with glycerol is shown in Figure 14.1. Three

Hydraulic Fracturing Chemicals and Fluids Technology. http://dx.doi.org/10.1016/B978-0-12-411491-3.00014-5
© 2013 Elsevier Inc. All rights reserved.

$$\begin{array}{l} H_2C{-}OH \\ HC{-}OH \\ H_2C{-}OH \end{array} \quad (HO)_3B \leftarrow\cdots \quad \begin{array}{l} HO{-}CH_2 \\ HO{-}CH \\ HO{-}CH_2 \end{array}$$

$$\downarrow$$

$$\left[\begin{array}{l} H_2C{-}O \\ HC{-}O \\ H_2C{-}OH \end{array} > B < \begin{array}{l} O{-}CH_2 \\ O{-}CH \\ HO{-}CH_2 \end{array} \right]^- H^+ + 2\,H_2O$$

FIGURE 14.1 Complexes of boric acid with glycerol.

hydroxyl units form an ester and one unit forms a complex bond. Here a proton will be released that lowers the pH. The scheme is valid also for polyhydroxy compounds. In this case, two polymer chains are connected via such a link.

The control of the delay time requires the control of the pH, the availability of borate ions, or both. Control of the pH can be effective in fresh water systems (Ainley et al., 1993). However, the control of borate is effective in both fresh water and sea water. This may be accomplished by using sparingly soluble borate species or by complexing the borate with a variety of organic species.

Borate crosslinked fracturing fluids have been successfully used in fracturing operations. These fluids provide excellent rheologic, fluid loss, and fracture conductivity properties over fluid temperatures up to 105 °C. The mechanism of borate crosslinking is an equilibrium process that can produce very high fluid viscosities under conditions of low shear (Cawiezel and Elbel, 1990).

A fracturing fluid containing borate is prepared in the following way (Harris et al., 1994):

Preparation 14.1. Introducing a polysaccharide polymer into (sea) water to produce a gel. Then an alkaline agent is added to the gel to obtain a pH of at least 9.5. Finally a borate crosslinking agent is added to the gel to crosslink the polymer. ■

A dry granular composition can be prepared in the following way (Harris and Heath, 1994):

Preparation 14.2. Dissolving from 0.2% to 1.0% of a water-soluble polysaccharide in an aqueous solution. Admixing a borate source with the aqueous gel formed before. Drying the thus formed borate crosslinked polysaccharide, and granulating the product. ■

A borate crosslinking agent can be boric acid, borax, an alkaline earth metal borate, or an alkali metal alkaline earth metal borate. The borate source, calculated as boric oxide, must be present in an amount of 5–30%.

Borated starch compositions are useful for controlling the rate of crosslinking of hydratable polymers in aqueous media for use in fracturing fluids. The borated starch compositions are prepared by reacting, in an aqueous medium, starch, and a borate source to form a borated starch complex. This complex provides a source of borate ions, which cause crosslinking of hydratable polymers in aqueous media (Sanner et al., 1996). Delayed crosslinking takes place at low temperatures.

Organic polyhydroxy compounds with hydroxyl moieties positioned in the *cis*-form on adjacent carbon atoms or on carbon atoms in a 1,3-relationship can react with borates to form five- or six-membered ring complexes. The reaction is fully reversible by changes in pH.

Depending on the concentration of the polymer and the borate anion, the crosslinking reaction may produce useful gels. Aqueous borate concentrates that provide a controllable crosslink time are highly appreciated. Sparingly soluble borate suspensions are suitable for hydraulic fracturing operations, since they adjust the time of crosslinking more consistently (Dobson et al., 2005). Examples of borate minerals are shown in Table 14.1.

The rheological characterization of borate crosslinked fluids using oscillatory measurements has been reported (Edy, 2010). It was demonstrated that the linear viscoelastic limit and the flow point frequency are dependent on temperature. The flow point frequency increases exponentially with the temperature.

The flow point is defined as the angular frequency at which the storage modulus G' and the loss modulus G'' are becoming equal. Here, a transition from elastic-dominated to viscous-dominated behavior will occur.

Titanium Compounds

Organic titanium compounds are useful as crosslinkers (Putzig and Smeltz, 1986; Putzig, 2010a). Aqueous titanium compositions often consist of mixtures of titanium compounds. Suitable organic titanium complexes are listed in Table 14.2.

The use of delay agents with titanium crosslinking agents has limited flexibility for use by the oil well service companies to stimulate or enhance recovery of oil or gas from a well or other subterranean formation since there are only a limited number of publications (Putzig, 2010a).

As delay agents, hydroxyalkylaminocarboxylic acids may be used. A preferred delay agent is bis(2-hydroxyethyl) glycine. This compound is shown in Figure 14.2.

TABLE 14.1 Sparingly Soluble Borate Minerals (Mondshine, 1986; Dobson et al., 2005; Parris and ElKholy, 2009)

Mineral	Formula
Probertite	$NaCaB_5O_9 \times 5\,H_2O$
Ulexite	$NaCaB_5O_9 \times 8\,H_2O$
Nobleite	$CaB_6O_{10} \times 4\,H_2O$
Gowerite	$CaB_6O_{10} \times 5\,H_2O$
Frolovite	$Ca_2B_4O_8 \times 7\,H_2O$
Colemanite	$Ca_2B_6O_{11} \times 5\,H_2O$
Meyerhofferite	$Ca_2B_6O_{11} \times 7\,H_2O$
Inyoite	$Ca_2B_6O_{11} \times 13\,H_2O$
Priceite	$Ca_4B_{10}O_{19} \times 7\,H_2O$
Tertschite	$Ca_4B_{10}O_{19} \times 20\,H_2O$
Ginorite	$Ca_2B_{14}O_{23} \times 8\,H_2O$
Pinnoite	$MgB_2O_4 \times 3\,H_2O$
Paternoite	$MgB_2O_{13} \times 4\,H_2O$
Kurnakovite	$Mg_1B_6O_{11} \times 15\,H_2O$
Inderite	$Mg_2B_6O_{11} \times 15\,H_2O$
Preobrazhenskite	$Mg_3B_{10}O_{18} \times 4\,H_2O$
Hydroboracite	$CaMgB_6O_{11} \times 6\,H_2O$
Inderborite	$CaMgB_6O_{11} \times 11\,H_2O$
Kaliborite (Heintzite)	$KMg_2B11O_{19} \times 9\,H_2O$
Veatchite	$SrB_6O_{10} \times 2\,H_2O$

TABLE 14.2 Organic Titanium Complexes (Putzig, 2010a)

Compound Type
Titanium alkanol amine complexes
Titanium diethanolamine complexes
Titanium triethanolamine complexes Titanium lactate
Titanium ethylene glycolate
Titanium acetylacetonate
Titanium ammonium lactate
Titanium diethanolamine lactate
Titanium triethanolamine lactate
Titanium diisopropylamine lactate
Titanium sodium lactate salts
Titanium sorbitol complexes

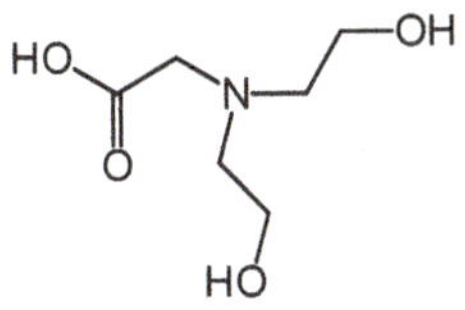

FIGURE 14.2 *N,N*-Bis(2-hydroxyethyl)glycine.

TABLE 14.3 Zirconium Compounds Suitable as Delayed Crosslinkers

Zirconium crosslinker/chelate	References
Hydroxyethyl-tris-(hydroxypropyl) ethylenediamine[a]	Putzig (1988)
Zirconium halide chelates	Ridland and Brown (1990)
Boron zirconium chelates[b]	Dawson and Le (1998); Sharif (1995)

[a] *Good high-temperature stability.*
[b] *High-temperature application, enhanced stability.*

Zirconium Compounds

Various zirconium compounds are used as delayed crosslinkers, c.f. Table 14.3. The initially formed complexes with low-molecular-weight compounds are exchanged with intermolecular polysaccharide complexes, which cause delayed crosslinking.

A diamine-based compound for complex forming is shown in Figure 14.3. Hydroxy acids are shown in Figure 14.4. Polyhydroxy compounds suitable for complex formation with zirconium compounds are shown in Figure 14.5.

Borozirconate complexes can be prepared by the reaction of tetra-*n*-propyl zirconate with triethanol amine and boric acid (Putzig, 2010b). The borozirconate complex can be used at a pH of 8–11.

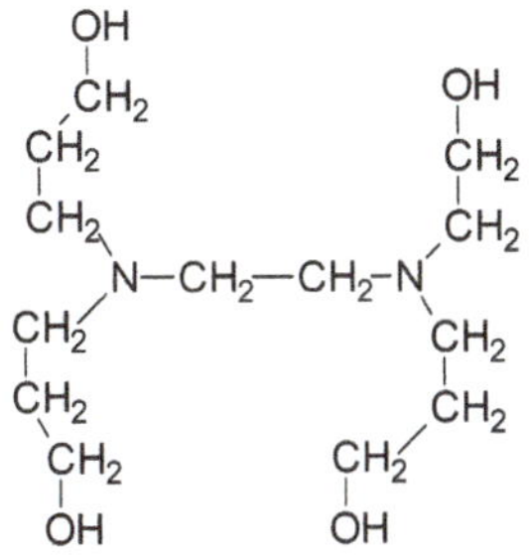

FIGURE 14.3 Hydroxyethyl-tris-(hydroxypropyl) ethylenediamine.

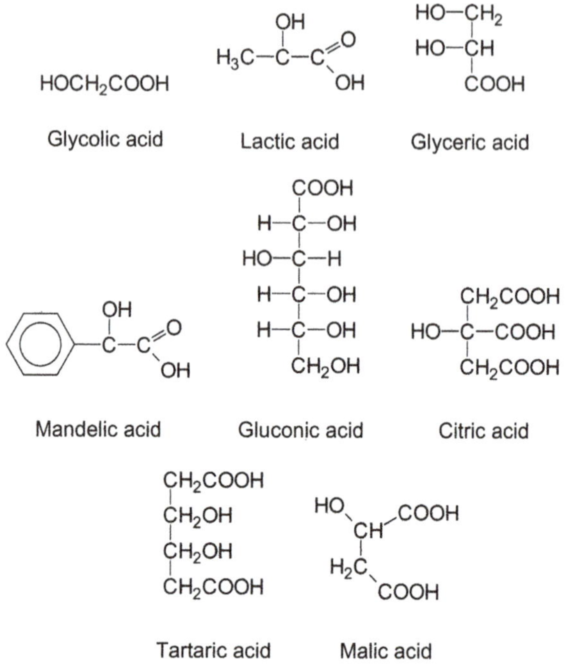

FIGURE 14.4 Hydroxy acids.

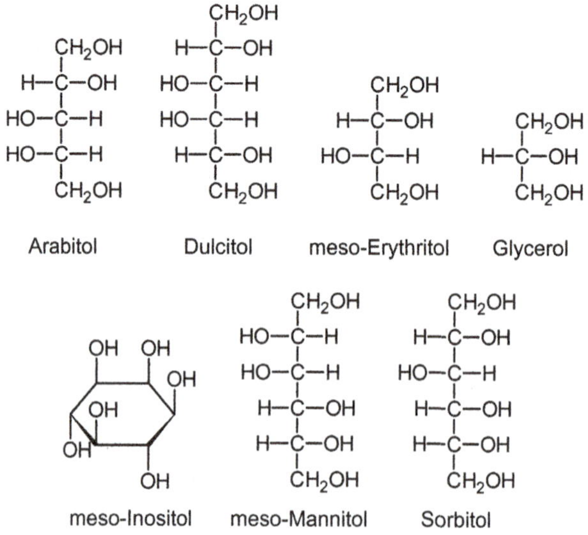

FIGURE 14.5 Polyalcohols for complex formation.

Guar

A hydrophobically modified guar gum can be used as an additive for drilling, completion, or servicing fluids (Audibert and Argillier, 1998). The modified gum is used together with polymers or reactive clay.

The grafting of an polyalkoxyalkyleneamide onto guar gum produces water-soluble guar derivatives (Bahamdan and Daly, 2007). The rheological properties of these products were investigated. The viscosity was measured at high temperatures and pressures in order to partially simulate the downhole conditions of oil wells. Treatment with zirconium lactate results in a better high-temperature stability and higher gel viscosities.

The viscosities of the crosslinked gels indicate that gels will be suitable to transport high amounts of proppant. In order to facilitate the removal of such gels from the formation, the hydrophobically modified guars were treated with an enzyme breaker system which produced fragments capable of producing stable emulsions when extracted with toluene. In this way, the cleanup process will be enhanced by the emulsification of the gel fragments produced by the hydrolysis of the gels (Bahamdan and Daly, 2007).

Innovations in guar and crosslinker technologies have resulted in the development of high-viscosity crosslinked borate fracturing fluids without increasing the polymer loadings. Low polymer borate fracturing fluids can be used successfully in various formations that have been previously believed to be too hot and/or too deep for low polymer fracturing fluids (Kostenuk and Gagnon, 2008).

Historically, polymer loadings of 3.6–4.2 $kg\,m^{-3}$ have been commonly pumped in the Western Canadian Sedimentary basin for formations deeper than 2500 m with bottomhole temperatures greater than 80 °C. These same formations are now fracture stimulated using the low polymer fluids with loadings of only 1.8 $kg\,m^{-3}$, thereby with remarkable results.

Low polymer fracture fluids can be used in place of fluids requiring higher polymer loadings with minimal changes to the overall design of the fracture treatment. A developed fluid composition can be pumped on-the-fly with conventional pump rates and proppant concentration because of its improved shear and temperature stability.

The advantages of a low polymer fracturing fluid include an increased production, lower treatment costs, and lower frictional pressure loss. In summary, low polymer fracturing fluids can be used in depths of up to 3250 m at temperatures grater than 100 °C (Kostenuk and Gagnon, 2008).

Hydroxypropyl Guar

Hydroxypropyl guar gum gel can be crosslinked with borates (Miller et al., 1996), titanates, or zirconates. Borate crosslinked fluids and linear hydroxyethyl cellulose gels are the most commonly used fluids for high-permeability fracture treatments. These gels are used for a hydraulic fracturing fluid under high temperature and high-shear stress.

Delayed Crosslinking Additives

Glyoxal (Dawson 1992a,b) is effective as a delay additive within a certain pH range. Glyoxal is shown in Figure 14.6. It bonds chemically with both boric acid and the borate ions to limit the number of borate ions initially available in solution for subsequent crosslinking of a hydratable polysaccharide (e.g., galactomannan).

The subsequent rate of crosslinking of the polysaccharide can be controlled by adjusting the pH of the solution. The mechanism of delayed crosslinking is shown in Figure 14.7. If two hydroxyl compounds with low molecular weight are exchanged with high-molecular-weight compounds, the hydroxyl units belonging to different molecules, then a crosslink is formed.

Other dialdehydes, keto aldehydes, hydroxyl aldehydes, ortho-substituted aromatic dialdehydes, and ortho-substituted aromatic hydroxyl aldehydes, have been claimed to be active in a similar way (Dawson, 1992a). Borate crosslinked guar fracturing fluids have been reformulated to allow the use at higher temperatures in both fresh water and sea water.

The temporary temperature range is extended for the use of magnesium oxide-delayed borate crosslinking of a galactomannan gum fracturing fluid by adding fluoride ions that precipitate insoluble magnesium fluoride (Nimerick et al., 1993).

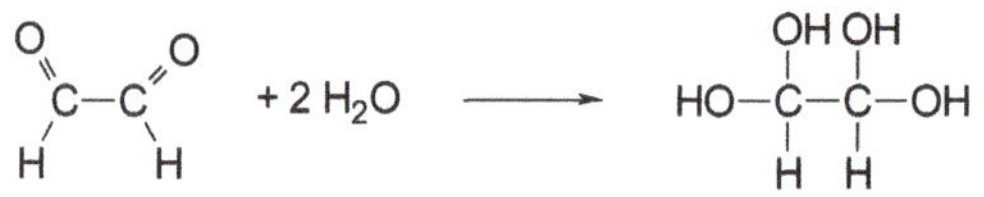

FIGURE 14.6 Glyoxal and hydrate formation.

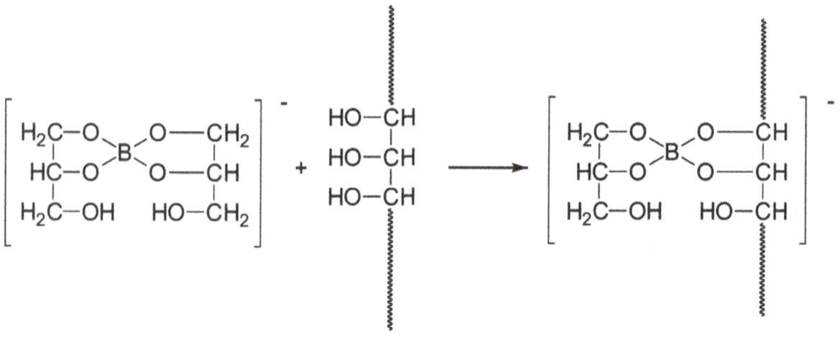

FIGURE 14.7 Delayed crosslinking.

Alternatively, a chelating agent for the magnesium ions may be added. With the precipitation of magnesium fluoride or the chelation of the magnesium ions, insoluble magnesium hydroxide cannot form at elevated temperatures, which would otherwise lower the pH and reverse the borate crosslinking reaction. The addition effectively extends the use of such fracturing fluids to temperatures of 135–150 °C.

Polyols, such as glycols or glycerol, can delay the crosslinking of borate in hydraulic fracturing fluids based on galactomannan gum (Ainley and McConnell, 1993). This is suitable for high-temperature applications up to 150 °C. In this case, low-molecular-weight borate complexes initially are formed but exchange slowly with the hydroxyl groups of the gum.

REFERENCES

Ainley, B. R., McConnell, S.B., 1993. Delayed borate crosslinked fracturing fluid. EP Patent 528 461, assigned to Pumptech NV and Dowell Schlumberger SA, 24 February 1993.

Ainley, B.R., Nimerick, K.H., Card, R.J., 1993. High-temperature, borate-crosslinked fracturing fluids: A comparison of delay methodology. In: Proceedings Volume. SPE Prod. Oper. Symp., Oklahoma, City, 21–23 March 1993, pp. 517–520.

Audibert, A., Argillier, J.F., 1998. Process and water-base fluid utilizing hydrophobically modified guars as filtrate reducers. US Patent 5 720 347, assigned to Inst. Francais Du Petrole, 24 February 1998.

Bahamdan, A., Daly, W.H., 2007. Hydrophobic guar gum derivatives prepared by controlled grafting processes-part II: rheological and degradation properties toward fracturing fluids applications. Polym. Adv. Technol. 18 (8), 660–672. http://dx.doi.org/10.1002/pat.875.

Barkat, O., 1987. Rheology of flowing, reacting systems: the crosslinking reaction of hydroxypropyl guar with titanium chelates. Ph.D. Thesis, Tulsa University.

Cawiezel, K.E., Elbel, J.L., 1990. A new system for controlling the crosslinking rate of borate fracturing fluids. In: Proceedings Volume. 60th Annual SPE California Reg Mtg., Ventura, California, 4–6 April 1990, pp. 547–552.

Dawson, J.C., 1992a. Method and composition for delaying the gellation (gelation) of borated galactomannans. US Patent 5 082 579, assigned to BJ Services Co., 21 January 1992.

Dawson, J.C., 1992b. Method for delaying the gellation of borated galactomannans with a delay additive such as glyoxal. US Patent 5 160 643, assigned to BJ Services Co., 3 November 1992.

Dawson, J.C., Le, H.V., 1998. Gelation additive for hydraulic fracturing fluids. US Patent 5 798 320, assigned to BJ Services Co., 25 August 1998.

Dobson Jr., J.W., Hayden, S.L., Hinojosa, B.E., 2005. Borate crosslinker suspensions with more consistent crosslink times. US Patent 6 936 575, assigned to Texas United Chemical Co., LLC., Houston, TX, 30 August 2005. <http://www.freepatentsonline.com/6936575.html>.

Edy, I.K.O., 2010. Rheological characterization of borate crosslinked fluids using oscillatory measurements. University of Stavanger, Stavanger, Norway, Master's Thesis.

Harris, P.C., Heath, S.J., 1994. Delayed release borate crosslinking agent. US Patent 5 372 732, assigned to Halliburton Co., Duncan, OK, 13 December 1994. <http://www.freepatentsonline.com/5372732.html>.

Harris, P.C., Norman, L.R., Hollenbeak, K.H., 1994. Borate crosslinked fracturing fluids. EP Patent 594 363, assigned to Halliburton Co., 27 April 1994.

Kostenuk, N., Gagnon, P., 2008. Polymer reduction leads to increased success: a comparative study. SPE Drill. Completion 23 (1), 55–60. http://dx.doi.org/10.2118/100467-PA.

Miller, II., W.K., Roberts, G.A., Carnell, S.J., 1996. Fracturing fluid loss and treatment design under high shear conditions in a partially depleted, moderate permeability gas reservoir. In: Proceedings Volume. SPE Asia Pacific Oil & Gas Conf., Adelaide, Australia, 28–31 October 1996, pp. 451–460.

Mondshine, T.C., 1986. Crosslinked fracturing fluids. US Patent 4619776, assigned to Texas United Chemical Corp. (Houston, TX), 28 October 1986. <http://www.freepatentsonline.com/4619776.html>.

Nimerick, K.H., Crown, C.W., McConnell, S.B., Ainley, B., 1993. Method of using borate crosslinked fracturing fluid having increased temperature range. US Patent 5259455, 9 November 1993.

Parris, M.D., ElKholy, I., 2009. Method and composition of preparing polymeric fracturing fluids. US Patent 7497263, assigned to Schlumberger Technology Corp., Sugar Land, TX, 3 March 2009. <http://www.freepatentsonline.com/7497263.html>.

Putzig, D.E., 1988. Zirconium chelates and their use for cross-linking. EP Patent 278684, assigned to Du Pont De Nemours & Co., 17 August 1988.

Putzig, D.E., 2010a. Cross-linking composition and method of use. US Patent 7732382, assigned to E.I. du Pont de Nemours and Co., Wilmington, DE, 8 June 2010. <http://www.freepatentsonline.com/7732382.html>.

Putzig, D.E., 2010b. Process to prepare borozirconate solution and use as cross-linker in hydraulic fracturing fluids. US Patent 7683011, 23 March 2010. <http://www.freepatentsonline.com/7683011.html>.

Putzig, D.E., Smeltz, K. C., 1986. Organic titanium compositions useful as cross-linkers. EP Patent 195531, 24 September 1986.

Ridland, J., Brown, D. A., 1990. Organo-metallic compounds. CA Patent 2002792, 16 June 1990.

Sanner, T., Kightlinger, A. P., Davis, J. R., 1996. Borate-starch compositions for use in oil field and other industrial applications. US Patent 5559082, assigned to Grain Processing Corp., 24 September 1996.

Sharif, S., 1995. Process for preparation of stable aqueous solutions of zirconium chelates. US Patent 5466846, assigned to Benchmark Res. & Technl In, 14 November 1995.

Chapter 15

Gel Stabilizers

Gel stabilizers are used to prevent the degradation of a crosslinked gel due to divalent or trivalent ion contamination. This is of particular importance for high-temperature fracturing jobs.

CHEMICALS

Suitable gel stabilizers are summarized in Table 15.1.

Sodium thiosulfate acts as an oxygen scavenger (Gupta and Carman, 2010a). Oxygen scavengers are reducing agents in that they remove dissolved oxygen from water by reducing molecular oxygen to compounds in which oxygen appears in the lower, i.e., −2 oxidation state. The reduced oxygen then combines with an acceptor atom, molecule, or ion to form an oxygen-containing compound. To be suitable as an oxygen scavenger, the reducing agent must have an exothermic heat of reaction with oxygen and must have reasonable reactivity at lower temperatures.

It has been suggested not to use thiosulfate-based high-temperature gel stabilizers. Instead, if high-temperature stability is desired or needed, it is suggested that instead triethanol amine should be used. Other suitable non-sulfur containing high-temperature gel stabilizers include methanol, diethanolamine, ethylenediamine, *n*-butylamine, and mixtures from these compounds (Crews, 2007).

Sodium gluconate and sodium glucoheptonate are commonly used for the chelation of cations such as calcium, magnesium, iron, manganese, and copper. They have been also found to be useful herein to complex titanate, zirconate, and borate ions for crosslink delay purposes. In addition, they are excellent iron complexing agents for enzyme breaker stability and crosslinked gel stability (Crews, 2006).

SPECIAL ISSUES

Water Softeners

A gel stabilizing effect is also provided by the addition of a water softener for a hard mix water (Le and Wood, 1993). A hard mix water means a mix of water

Hydraulic Fracturing Chemicals and Fluids Technology. http://dx.doi.org/10.1016/B978-0-12-411491-3.00015-7
© 2013 Elsevier Inc. All rights reserved.

TABLE 15.1 Gel Stabilizers

Compound	References
Sodium thiosulfate	Dawson and Le (1998), Willberg and Nagl (2004), Lord et al. (2005)
Sodium gluconate	Crews (2006)
Sodium glucoheptonate	Crews (2006)
Diethanolamine	Crews (2006)
Triethanolamine	Crews (2006)
Methanol	Crews (2006)
Hydroxyethylglycine	Crews (2006)
Tetraethylenepentamine	Crews (2006)
Ethylenediamine	Crews (2006)

with an excess of about 1000ppm total dissolved solids in terms of $CaCO_3$ equivalents.

The usual field conditions encountered will involve a hard mix water with 3000–7000ppm of total dissolved solids. A gel stabilizing water softener will bring down the free, uncomplexed multivalent ions to concentrations less than 3000ppm of $CaCO_3$ equivalents.

Preferred water softeners operate as chelating or sequestering agents and are selected from the group consisting of salts of inorganic polyphosphates, amino polycarboxylic acids such as ethylenediaminetetraacetic acid, poly-(acrylate)s, salts of typical phosphonate scale inhibitors, e.g., diethylenetriaminepenta(methylene phosphonic acid), salts of nitrilotrimethylenephosphonic acid, ethylenediamine tetramethylene phosphonic acid and ethylenediamine hydroxy diphosphonic acid. A particularly preferred water softener is the sodium salt of diethylenetriaminepenta(methylene phosphonic acid) (Le and Wood, 1993).

Borate Reserve

Borated guar systems have been employed with either slow dissolving metal oxides, which slowly increase the fluid alkalinity, which in turn promotes the crosslinking reaction, or have directly used calcium borate salts with a poor solubility in water. Additionally, the use of sparingly soluble borates achieves some degree of enhanced thermal stability of the gels since a reserve of boron is available for crosslinking over an extended period of time (Mondshine, 1986).

A complexing agent for controlled delay and improved high-temperature gel stability of borated fracturing fluids has been developed (Dawson, 1992). A base fluid is first prepared by blending together an aqueous fluid and a hydratable polymer which is capable of gelling in the presence of borate ions.

The complexing agent is prepared by mixing a crosslinking additive capable of furnishing borate ions in solution with a delay additive.

The delay additive is effective, within a selected pH range, to chemically bond with both boric acid and the borate ions produced by the crosslinking additive to thereby limit the number of borate ions initially available in solution for subsequent crosslinking of the hydratable polysaccharide. In addition to providing a more precise control of the delay time, the complexing agent provides a reserve of borate which provides improved gel stability at higher temperatures. In detail the method of preparation has been given as follows (Dawson, 1992):

Preparation 15.1. Into 300 parts of 40% aqueous glyoxal are added, with stirring, 130 parts of sodium borate decahydrate yielding a milky white suspension. Then, 65 parts of 25% aqueous yellow solution. The solution pH can range from 4.90 to 6.50. Afterward, 71.4 parts of 70% aqueous sorbitol are added to the solution followed by heating to 95 °C for 3 h. During heating, the solution color changes from pale yellow to amber. After cooling to ambient temperature, the solution pH ranges between 4.50 and 5.00. Each gallon of complexing agent contains a boron concentration equivalent to 0.29 lb of elemental boron or 1.65 lb of boric acid. ■

The complexing agent can also be used in a dual crosslink system. Thus, by mixing a traditional borate crosslinking agent such as boric acid or sodium borate with the complexing agent, a faster crosslinking time is observed. This effect can be used to enhance the early performance of the system at high temperatures by adding a small amount of either boric acid or sodium borate. This small amount of traditional crosslinking agent will give extra viscosity to the fluid as it is transporting sand through the tubing string from the well surface. The small increase in viscosity which is observed does not otherwise interfere with the desirable properties of the fluid (Dawson, 1992).

Electron Donor Compounds

Compositions for the reduction of the thermal degradation of aqueous gels by the addition of a gel stabilizer using an oxime as electron donor compound have been reported (Pakulski and Gupta, 1994). Such donor compounds are capable of stabilizing gels at temperatures as high as 150 °C.

It has been discovered that chalcogen heterocyclic compounds containing oxygen or sulfur are useful for extending the high-temperature effectiveness of aqueous gels commonly used in oilfield operations. It has been found that such compounds prevent the thermal degradation of such gels at temperatures as high as 205 °C. The kinetics of reaction is fast enough even at low temperatures to stabilize such gels. The sterically unhindered oxygen atom on the heterocyclic compound carries two unshared pairs of electrons which provide the electron donation for gel stability (Gupta and Carman, 2010b).

TABLE 15.2 Heterocyclic Electron Donating Compounds (Gupta and Carman, 2010a)

Compound	Compound
Tetrahydrofuran	3-Methyltetrahydrofuran
2-(Diethoxymethyl)furan	2-Methyl-5-(methylthio)furan
Difurfurylsulfide	Dibenzofuran
1,2,3,4-Tetrahydrodibenzofuran	2-Acetyl-5-methylfuran
Tetrahydrofurfuryl bromide	2,3-Dihydrofuran
2,2-Dimethyltetrahydrofuran	2,5-Dimethyltetrahydrofuran
2,3,4,5-Tetramethylfuran	2-Methyl-5-propionyl-furan
3-Acetyl-2,5-dimethylfuran	2-Acetylfuran
2-Acetyl-2,5,-dimethylfuran	Bromotrichlorodibenzofuran
Thiophene	Succinic anhydride
Maleic anhydride	

Chalcogen heterocyclic electron donating compounds are summarized in Table 15.2.

The chalcogen heterocyclic electron donating compound may be used in combination with a second conventional oxygen scavenger (Gupta and Carman, 2010a). Examples of such compounds are summarized in Table 15.3 and mixtures thereof. Typically, when employed, the weight percent ratio of chalcogen heterocyclic electron donating compound: second conventional oxygen scavenger is from about 0.01 to 1, preferably from about 0.1 to 0.2.

The chalcogen heterocyclic electron donating compound can be added to the mix water prior to or at the same time as other additives are added to a water-based oilfield gel. The chalcogen heterocyclic electron donating compound could be added on-the-fly if necessary for continuous process operation (Gupta and Carman, 2010b).

TABLE 15.3 Oxygen Scavengers

Compound	Compound	Compound
Sodium thiosulfate	Sodium sulfite	Sodium bisulfite
Pyrogallic acid	Pyrogallol	Catechol
Sodium erythrobate	Ascorbic acid	Resorcinol
Stannous chloride	Quinone	Hydroquinone

Effects of pH on Gel Stability

Standard hydrocarbon gelling systems were prepared and varying quantities of acid or base were added. An aluminum-based system was prepared by using 3.12 m mol of aluminum chlorohydrate as a 50% aqueous solution with 1.8 ml of Rhodafac™ LO-11A-LA in 300 ml diesel. Iron gels were prepared by using 6.75 m mol of ferric sulfate as a 50% aqueous solution with 1.8 ml of Rhodafac™ LO-11A-LA in 300 ml diesel.

It is apparent in both systems that the addition of acid has a negative impact on both the rheology and the temperature stability, whereas the addition of a base has a beneficial effect on the rheology. The iron system was found to be more tolerant to the addition of acids or bases (Lawrence and Warrender, 2010).

Tradenames in References

Tradename	Supplier
Description	
Alkaquat™ DMB-451	Rhodia Canada Inc.
Dimethyl benzyl alkyl ammonium chloride (Lawrence and Warrender, 2010)	
Geltone® (Series)	Halliburton Energy Services, Inc.
Organophilic clay (Lawrence and Warrender, 2010)	
Rhodafac® LO-11A-LA	Rhodia Inc. Corp.
Phosphate ester (Lawrence and Warrender, 2010)	

REFERENCES

Crews, J.B., 2006. Biodegradable chelant compositions for fracturing fluid. US Patent 7 078 370, assigned to Baker Hughes Inc., Houston, TX, 18 July 2006. <http://www.freepatentsonline.com/7078370.html>.

Crews, J.B., 2007. Fracturing fluids for delayed flow back operations. US Patent 7 256 160, assigned to Baker Hughes Inc., Houston, TX, 14 August 2007. <http://www.freepatentsonline.com/7256160.html>.

Dawson, J.C., 1992. Method for improving the high temperature gel stability of borated galactomannans. US Patent 5 145 590, assigned to BJ Services Co., Houston, TX, 8 September 1992. <http://www.freepatentsonline.com/5145590.html>.

Dawson, J.C., Le, H.V., 1998. Gelation additive for hydraulic fracturing fluids. US Patent 5 798 320, assigned to BJ Services Co., 25 August 1998.

Gupta, D.V.S., Carman, P.S., 2010a. Method of treating a well with a gel stabilizer. US Patent 7 767 630, assigned to BJ Services Co., LLC, Houston, TX, 3 August 2010. <http://www.freepatentsonline.com/7767630.html>.

Gupta, D.V.S., Carman, P.S., 2010b. Method of treating a well with a gel stabilizer. US Patent Application 20100016182, 21 January 2010. <http://www.freepatentsonline.com/20100016182.html>.

Lawrence, S., Warrender, N., 2010. Crosslinking composition for fracturing fluids. US Patent 7 749 946, assigned to Sanjel Corp., Calgary, Alberta, CA, 6 July 2010. <http://www.freepatentsonline.com/7749946.html>.

Le, H.V., Wood, W.R., 1993. Method for increasing the stability of water-based fracturing fluids. US Patent 5 226 481, assigned to BJ Services Co., Houston, TX, 13 July 1993. <http://www.freepatentsonline.com/5226481.html>.

Lord, P.D., Terracina, J., Slabaugh, B., 2005. High temperature seawater-based cross-linked fracturing fluids and methods. US Patent 6 911 419, assigned to Halliburton Energy Services, Inc., Duncan, OK, 28 June 2005. <http://www.freepatentsonline.com/6911419.html>.

Mondshine, T.C., 1986. Crosslinked fracturing fluids. US Patent 4 619 776, assigned to Texas United Chemical Corp., Houston, TX, 28 October 1986. <http://www.freepatentsonline.com/4619776.html>.

Pakulski, M.K., Gupta, D.V.S., 1994. High temperature gel stabilizer for fracturing fluids. US Patent 5 362 408, assigned to The Western Company of North America, Houston, TX, 8 November 1994. <http://www.freepatentsonline.com/5362408.html>.

Willberg, D., Nagl, M., 2004. Method for preparing improved high temperature fracturing fluids. US Patent 6 820 694, assigned to Schlumberger Technology Corp., Sugar Land, TX, 23 November 2004. <http://www.freepatentsonline.com/6820694.html>.

Chapter 16

Gel Breakers

After a formation is adequately fractured and the proppant is in place, the fracturing fluid is recovered typically through the use of breakers. Breakers generally reduce the fluid's viscosity to a low enough value that allows the proppant to settle into the fracture and thereby increase the exposure of the formation to the well. Breakers work by reducing the molecular weight of the polymers, i.e., degrading the polymer. The fracture then becomes a high-permeability conduit for fluids and gas to be produced back to the well (Armstrong, 2012).

Besides providing a breaking mechanism for the gelled fluid to facilitate recovery of the fluid, breakers can also be used to control the timing of the breaking of the fracturing fluids, which is important. Gels that break prematurely can cause suspended proppant material to settle out of the fluid before being introduced a sufficient distance into the produced fracture. Premature breaking can also result in a premature reduction in the fluid viscosity resulting in a less than desirable fracture length in the fracture being created (Armstrong, 2012).

On the other hand, gelled fluids that break too slowly can cause slow recovery of the fracturing fluid and a delay in resuming the production of formation fluids. Additional problems can result, such as the tendency of proppant to become dislodged from the fracture, resulting in a less than desirable closing and decreased efficiency of the fracturing operation.

Optimally, the fracturing gel will begin to break when the pumping operations are concluded. For practical purposes, the gel should be completely broken within a specific period of time after completion of the fracturing period. At higher temperatures, for example, about 24 h are sufficient. A completely broken gel will be taken to mean one that can be flushed from the formation by the flowing formation fluids or that can be recovered by a swabbing operation (Armstrong, 2012).

GEL BREAKING IN WATER-BASED SYSTEMS

In general, there are two methods for combining the fracturing fluid and the breaker (Carpenter, 2009):

Hydraulic Fracturing Chemicals and Fluids Technology. http://dx.doi.org/10.1016/B978-0-12-411491-3.00016-9
© 2013 Elsevier Inc. All rights reserved.

1. Mixing the breaker with the fracturing fluid prior to sending the fracturing fluid downhole, or
2. Sending the fracturing downhole, and afterwards the breaker.

The first method is favored at least because of convenience. It is easier to mix the fluids at the surface and send the mixture downhole. A disadvantage of this blending method is that the breaker can decrease the viscosity of the fracturing fluid before the desired time.

In the second method, the fracturing fluid is sent downhole, and the breaker is sent downhole later. While sending the breaker downhole later is inconvenient, in this method the breaker does not decrease the viscosity of the fracturing fluid prematurely (Carpenter, 2009).

After the fracturing job, the properties of the formation should be restored. Maximal well production can be achieved only when the solution viscosity and the molecular weight of the gelling agent are significantly reduced after the treatment, that is, the fluid is degraded.

Comprehensive research on the degradation kinetics of a hydroxypropyl guar fracturing fluid by enzyme, oxidative, and catalyzed oxidative breakers was performed (Craig, 1991; Craig and Holditch, 1993a,b). Changes in viscosity were measured as a function of time.

The studies revealed that enzyme breakers are effective only in acid media at temperatures of 60 °C or below. In an alkaline medium and at temperatures below 50 °C, a catalyzed oxidative breaker system was the most effective breaker. At temperatures of 50 °C or higher, hydroxypropyl guar fracturing fluids can be degraded by an oxidative breaker without a catalyst.

OXIDATIVE BREAKERS

Among the oxidative breakers, alkali, metal hypochlorites, and inorganic and organic peroxides have been described in the literature (Bielewicz and Kraj, 1998). These materials degrade the polymer chains by oxidative mechanisms. Carboxymethyl cellulose, guar gum, or partially hydrolyzed poly(acrylamide)s were used for testing a series of oxidative gel breakers in a laboratory study.

Hypochlorite Salts

Hypochlorites are powerful oxidants and therefore may degrade polymeric chains. They are often used in combination with tertiary amines (Williams et al., 1987). The combination of the salt and the tertiary amine increases the reaction rate more than the application of a hypochlorite alone. A tertiary amino galactomannan may serve as an amine source (Langemeier et al., 1989).

This also serves as a thickener before breaking. Hypochlorites are also effective for breaking stabilized fluids (Walker and Shuchart, 1995). Sodium thiosulfate has been proposed as a stabilizer for high-temperature applications.

Peroxide Breakers

Alkaline earth metal peroxides have been described as delayed gel breakers in alkaline aqueous fluids containing hydroxypropyl guar (Mondshine, 1993). The peroxides are activated by increasing the temperature of the fluid.

It has been shown that calcium peroxide is more effective than magnesium peroxide in breaking down the polymer at lower temperatures. Magnesium peroxide can be added to a fluid containing a polysaccharide polymer to produce a delayed break at relatively high temperatures, whereas calcium peroxide can be added to produce a delayed break at relatively low temperatures. Perphosphate esters or amides can be used for oxidative gel breaking (Laramay et al., 1995). Whereas the salts of the perphosphate ion interfere with the action of the crosslinkers, the esters and amides of perphosphate do not.

Fracturing fluids that contain these breakers are useful for fracturing deeper wells operating at temperatures of 90–120 °C and using metal ion crosslinkers, such as titanium and zirconium. Breaker systems based on persulfates have also been described (Harms, 1992). In addition, organic peroxides are suitable for gel breaking (Dawson and Le, 1995). The peroxides need not be completely soluble in water. The time needed to break is controlled in the range of 4–24 h by adjusting the amount of breaker added to the fluid.

Redox Gel Breakers

Basically, gel breakers act according to a redox reaction. Copper(II) ions and amines can degrade various polysaccharides (Shuchart et al., 1999).

DELAYED RELEASE OF ACID

Regained permeability studies with hydroxyethyl cellulose polymer in high-permeability cores revealed that persulfate-type oxidizing breakers and enzyme breakers do not adequately degrade the polymer. Sodium persulfate breakers were found to be thermally decomposed, and the decomposition was accelerated by minerals present in the formation.

The enzyme breaker adsorbed onto the formation but still partly functioned as a breaker. Dynamic fluid loss tests at reduced pH with borate crosslinked gels suggest that accelerated leakoff away from the wellbore could be obtained through the use of a delayed release acid. Rheologic measurements confirmed that a soluble delayed release acid could be used to convert a borate crosslinked fluid into a linear gel (Noran et al., 1995).

Hydroxyacetic Acid Condensates

A condensation product of hydroxyacetic acid can be used as a fluid loss material in a fracturing fluid in which another hydrolyzable aqueous gel is used

$$\sim O-H_2C-C(=O)-O-H_2C-C(=O)\sim \xrightarrow{H_2O} HO-H_2C-C(=O)-OH$$

FIGURE 16.1 Hydrolysis of poly(glycolic acid).

(Cantu and Boyd, 1989; Cantu et al., 1990a,b). The hydroxyacetic acid condensation product degrades at formation conditions to set free hydroxyacetic acid, which breaks the aqueous gel. This mechanism may be used for delayed gel breaking, as shown in Figure 16.1. Here the permeability is restored without the need for separate addition of a gel breaker, and the condensation product acts a fluid loss additive.

ENZYME GEL BREAKERS

Enzymes are catalytic and substrate specific and will catalyze the hydrolysis of specific bonds on the polymer. Using enzymes for controlled breaks circumvents the oxidant temperature problems, as the enzymes are effective at the lower temperatures. An enzyme will degrade many polymer bonds in the course of its useful lifetime. Unfortunately, enzymes operate under a narrow pH range and their functional states are often inactivated at high pH values.

Conventional enzymes used to degrade galactomannans have maximum catalytic activities under mildly acidic to neutral conditions, i.e., a pH of 5–7 (Armstrong, 2012). Extreme temperature stable and polymer-specific enzymes have been developed (Brannon and Tjon-Joe-Pin, 1994; Sarwar et al., 2011; Jihua and Sui, 2011).

Basic studies have been performed to investigate the performance of enzymes. The products of degradation, the kinetics of degradation, and limits of application, such as temperature and pH, have been analyzed (Slodki and Cadmus, 1991; Craig et al., 1992). Because enzymes degrade chemical linkages highly selectively, no general-purpose enzyme exists, but for each thickener, a selected enzyme must be applied to guarantee success. Enzymes suitable for particulate systems are shown in Table 16.1.

Enzymes are suitable to break the chains of the thickener directly. Other systems also have been described that enzymatically degrade polymers, which

TABLE 16.1 Polymer Enzyme Systems

Polymer	References
Xanthan[a]	Ahlgren (1993)
Mannan-containing hemicellulose[b]	Fodge et al. (1996).

[a]*Elevated temperatures and salt concentrations.*
[b]*High alkalinity and elevated temperature.*

degrade into organic acid molecules. These molecules are actually active in the degradation of the thickener (Harris and Hodgson, 1998).

Interactions

Despite their advantages over conventional oxidative breakers, enzyme breakers have limitations because of interferences and incompatibilities with other additives. Interactions between enzyme breakers and fracturing fluid additives including biocides, clay stabilizers, and certain types of resin coated proppants have been reported (Prasek, 1996).

ENCAPSULATED GEL BREAKERS

The breaker chemical in encapsulated gel breakers is encapsulated in a membrane that is not permeable or is only slightly permeable to the breaker. Therefore, the breaker may not come in contact initially with the polymer to be degraded. Only with time the breaker diffuses out from the capsulation, or the capsulation is destroyed so that the breaker can act successfully.

Encapsulated gel breakers find a wide field of application for delayed gel breaking. The breaker is prepared by encapsulating it with a water-resistant coating. The coating shields the fluid from the breaker so that a high concentration of breaker can be added to the fluid without causing premature loss of fluid properties, such as viscosity or fluid loss control.

Critical factors in the design of encapsulated breakers are the barrier properties of the coating, release mechanisms, and the properties of the reactive chemicals. For example, a hydrolytically degradable polymer can be used as the membrane (Muir and Irwin, 1999).

This method of delayed gel breaking has been reported both for oxidative breaking and for enzyme gel breaking. Formulations of encapsulated gel breakers are shown in Table 16.2. Membranes for encapsulators are shown in Table 16.3.

TABLE 16.2 Use of Encapsulation in Delayed Gel Breaking

Breaker System	References
Ammonium persulfate[a]	Gulbis et al. (1990a,b,1992,); King et al. (1990)
Enzyme breaker[b]	Gupta and Prasek (1995)
Complexing agents[c]	Boles et al. (1996)

[a]*Guar or cellulose derivatives.*
[b]*Open cellular coating.*
[c]*For titanium and zirconium; wood resin encapsulated.*

TABLE 16.3 Membranes for Encapsulated Breakers

Membrane Material	References
Poly(amide)[a]	Gupta and Cooney (1992)
Crosslinked elastomer	Manalastas et al. (1992)
Partially hydrolyzed acrylics crosslinked with aziridine prepolymer or carbodiimide[b]	Hunt et al. (1997); Norman and Laramay (1994); Norman et al. (2001)
Seven percent Asphalt and 93% neutralized sulfonated ionomer	Swarup et al. (1996)

[a]*For peroxide particle sizes 50–420 μm.*
[b]*Enzyme coated on cellulose derivative.*

GEL BREAKING OF GUAR

Guar-based polymer gels are used in the oil and gas industry to viscosity fluids used in hydraulic fracturing of hydrocarbon production wells (Johnson et al., 2011). After fracturing, the gel and filter cake must be degraded to obtain high hydraulic conductivity of the fracture and rock/fracture interface. Enzymes are widely used to achieve this but a high concentration may result in a premature degradation, or conversely in a failure to gel. Also the denaturation of enzymes at harsh pH and temperatures limits their applicability.

Maximal well production can be achieved only when the solution viscosity and the molecular weight of the gelling agent are significantly reduced after the treatment. However, the reduction of the fracturing fluid viscosity, the traditional method of evaluating these materials, does not necessarily indicate that the gelling agent has been thoroughly degraded also.

The reaction between hydroxypropyl guar and the oxidizing agent (ammonium peroxydisulfate) in an aqueous potassium chloride solution was studied (Hawkins, 1986) under controlled conditions to determine changes in solution viscosity and the weight average of the molecular mass of hydroxypropyl guar.

Bromine compositions used for gel breaking can be stabilized with sodium sulfamate (Carpenter, 2009). The sulfamate used in the production of such breakers is effective in stabilizing the active bromine species over long periods of time, especially at a pH of 13. For example, a WELLGUARD™ 7137 gel breaker is stable for greater than one year if protected from sunlight. The halogen source of the breaker are interhalogen compounds, bromine chloride, or mixtures of bromine and chlorine.

Unlike hypobromites (^{-}OBr), these type breakers do not oxidize or otherwise destroy organic phosphonates that are typically used as corrosion and

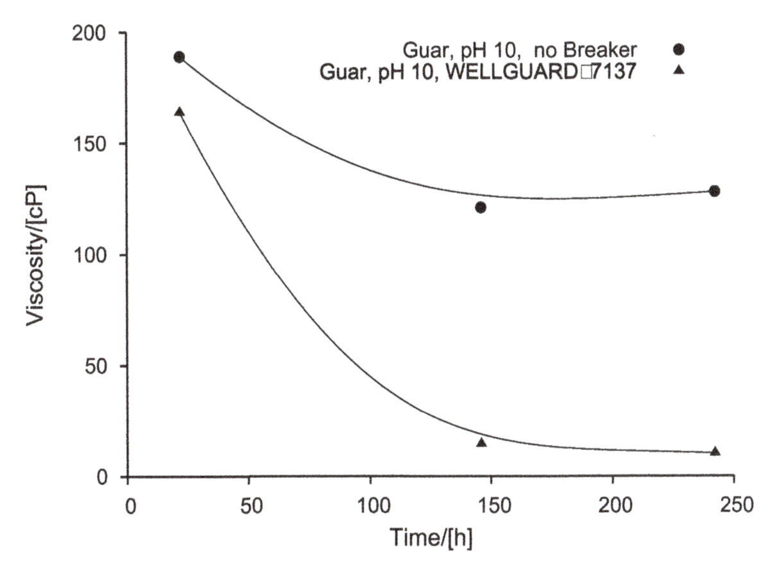

FIGURE 16.2 Effect of halogen-based breakers on guar (Carpenter, 2009).

TABLE 16.4 EDTA-Related Gel Breakers (Crews, 2007a)

Complex Compound
Tetrasodium propylenediamine tetraacetic acid
Trisodium hydroxyethylenediamine tetraacetic acid
Trisodium nitrilotriacetic acid
Trisodium ethylenediaminetriacetic acid
Disodium ethylenediamine diacetic acid
Disodium calcium dihydrate ethylenediamine diacetic acid
Tetraammonium ethylenediamine tetraacetic acid

scale inhibitors. Further, the breakers exhibit a low corrosivity against metals, especially against ferrous alloys. This is the result of the low oxidation-reduction potential of these breakers (Carpenter, 2007, 2009). The effect of the breakers on guar is shown in Figure 16.2. The composition was prepared and studied at 50 °C (120 °F).

Borate crosslinked guar polymer gels can be broken with ethylenediamine tetraacetic acid (EDTA) compounds (Crews, 2007a). EDTA and other amino carboxylic acid compounds can break the gelled fracturing fluid. Examples are shown in Table 16.4.

It is believed these breakers act directly on the polymer itself and not on any crosslinker that may be present. Also polyhydroxy compounds can break guar gels, and moreover gels formed by polysaccharides. These polyhydroxy compounds include mannitol and sorbitol. The polyols can be used in combination of enzyme breakers (Crews, 2007b).

Enzyme Breaking of Guar

Because most guar polymers are crosslinked at pH values of 9.5–11 for fracturing applications, a need exists for a breaker that can degrade guar-based fracturing fluids within this range of pH. Glycoside hydrolases have been developed for these purposes (Armstrong, 2012).

Enzymes of the subfamily 8 of glycoside hydrolase may be used. Such retaining enzymes include the enzyme breaker derived from a gene of the alkaliphilic Bacillus species N16-5. These enzymes exhibit a pH optimum of enzymatic activity. The preparation of such β-mannanase enzymes has been described (Ma et al., 2004).

The enzyme is prepared from culture broth of alkaliphilic Bacillus sp. N16-5. The optimum activity is pH 9.5 and 70 °C. The enzyme is composed of a single polypeptide chain with a molecular weight of 55 kDa. The enzyme is effective in hydrolyzing galactomannan and glucomannan, into a series of oligosaccharides and monosaccharides.

For use for fracturing purposes it is important that such an enzyme breaker is catalytically active and temperature stable in a temperature range of about 15–107 °C (Tjon-Joe-Pin, 1993).

Because the enzyme breaker has a maximum activity under alkaline pH ranges, it can be combined with other breakers that operate in different pH ranges to allow for better control of hydrolysis of fracturing fluids over a much greater pH range. The crosslinked polymer gel can further include a second enzyme breaker that is catalytically active and temperature stable in a pH range of 4–8. Suitable enzymes that can be used have been described (Tjon-Joe-Pin, 1993).

Divalent cations can have beneficial or detrimental effects on the activity of the enzyme breaker. Mg^{2+} increases the activity, while the presence of Co^{2+} decreases the enzyme activity (Tjon-Joe-Pin, 1993).

Nanoparticles can be made with different ratios of poly(ether imide) (PEI) and dextran sulfate (Cordova et al., 2008). The self-assembly of PEI and dextran sulfate results in the formation of 100–200 nm particles that can efficiently entrap chromium while maintaining their colloidal stability in water or gelant solution. The addition of chromium chloride to the composition typically produces gels within minutes, however, chromium is efficiently sequestered in the suspensions of the polyelectrolyte complexes. This effects a significant delay in gel formation. The chemical structure of a borate crosslinked guar is shown in Figure 16.3.

The ether bonds formed are vulnerable to a cleavage by an enzyme, such as pectinase (Barati et al., 2011). On the other hand, PEI-dextran sulfate

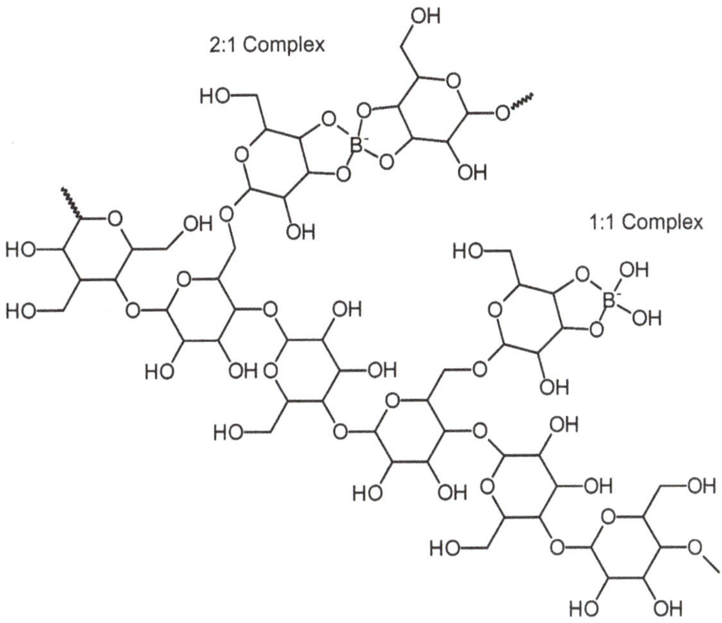

FIGURE 16.3 Borate crosslinked guar (Barati et al., 2011).

polyelectrolyte complexes were used to entrap guar-degrading enzymes and so to obtain a delayed release and to protect the enzyme from harsh conditions.

A degradation caused by viscoelastic properties of borate crosslinked hydroxypropyl guar gel by commercial enzyme loaded in polyelectrolyte nanoparticles was delayed up to 11 h, in comparison to only about 3 h for equivalent systems in those the enzyme mixture was not entrapped. In addition, the PEI-dextran nanoparticles protect the enzymes from denaturation at elevated temperatures and pH (Johnson et al., 2011; Barati et al., 2011, 2012).

Oxidative and enzyme breakers have been used to study the degradation of guar polymer gel as a function of time, temperature, and breaker concentration itself (Kyaw et al., 2012).

Chemical breakers were added to reduce the viscosity of the guar polymer by cleaving the polymer into fragments. The reduction of the viscosity facilitates the flowback of the residual polymer. In this way, a rapid recovery of the polymer from the proppant pack is achieved. The use of ineffective breakers or a wrong application of a breaker can result in screen-outs or flowback of viscous fluids. Both can significantly decrease the productivity of the well. The activity of the breakers at low and medium temperatures was evaluated.

Several methods of testing have been described in some detail, among others, a crosslink test and a break time test (Kyaw et al., 2012).

The residue after break test may determine the amount of unbroken gel after the breaking procedure ends. This is basically a filtration where the percentage of residue of unbroken polymer is measured (Sarwar et al., 2011).

VISCOELASTIC SURFACTANT GELLED FLUIDS

The viscosity of fluids that are viscosified with viscoelastic surfactants (VESs) can be controlled by fatty acid salts. For example, a brine fluid gelled with an amine oxide surfactant may have its viscosity broken with a composition containing naturally occurring fatty acid salts from canola oil or corn oil (Crews, 2010).

The alteration of the fatty acid or the saponification reaction may occur during mixing and pumping of the fluid downhole. The method may also be used where most of the saponification occurs within the reservoir shortly after the treatment is over. Alternatively, the components may be preformed and added later as an external breaker solution to remove the VES-gelled fluids that have been already placed downhole.

It may be possible that first an increase in viscosity and afterwards a decrease in viscosity may occur. When canola oil is saponified with CaOH, initially a slight increase of the VES fluid is observed, followed by a breaking reaction (Crews, 2010). The increase in viscosity is explained as the initially particular saponified fatty acids may act as viscosity enhancing cosurfactants for the fluid containing VESs.

GRANULES

Granules may also be helpful in delayed breaking. Granules with 40–90% of sodium or ammonium persulfate breaker and 10–60% of an inorganic powdered binder, such as clay, have been described (McDougall et al., 1993). The granules exhibit a delayed release of the breaker.

Other chemicals acting as delayed gel breakers are also addressed as controlled solubility compounds or cleanup additives that slowly release certain salts. Poly(phosphate)s have been described for such polymers (Mitchell et al., 2001).

Granules composed of a particulate breaker dispersed in a wax matrix are used in fracturing operations to break hydrocarbon liquids gelled with salts of alkyl phosphate esters. The wax granules are solid at surface temperature and melt or disperse in the hydrocarbon liquid at formation temperature, releasing the breaker to react with the gelling agent (Acker and Malekahmadi, 2001).

Gel Breakers for Oil-Based Systems

Gel breakers used in nonaqueous systems have a completely different chemistry than those used in aqueous systems. A mixture of hydrated lime and sodium bicarbonate is useful in breaking nonaqueous gels (Syrinek and Lyon, 1989). Sodium bicarbonate used by itself is totally ineffective for breaking the fracturing fluid for aluminum phosphate-based or aluminum phosphate ester-based gellants. An alternative is to use sodium acetate as a gel breaker for nonaqueous gels.

Enzyme-Based Gel Breaking

The viscosities of fluids viscosified with VES may be reduced by the direct or indirect action of a biochemical agent, such as bacteria, fungi, or enzymes. The biochemical agent may directly attack the VES itself, or some other component in the fluid that produces a byproduct that then causes viscosity reduction. The biochemical agent may disaggregate or otherwise attack the micellar structure of the VES-gelled fluid. The biochemical agent may produce an enzyme that reduces viscosity by one of these mechanisms.

A single biochemical agent may operate simultaneously by two different mechanisms, such as by degrading the VES directly, as well as another component, such as a glycol, the latter mechanism in turn producing a byproduct, e.g., an alcohol that causes viscosity reduction.

Alternatively, two or more different biochemical agents may be used simultaneously. In a specific instance, a brine fluid gelled with an amine oxide surfactant can have its viscosity broken with bacteria such as *Enterobacter cloacae*, *Pseudomonas fluorescens*, or *Pseudomonas aeruginosa* (Crews, 2006).

Breaker Enhancers for VES

Oil-soluble surfactants may be used as breaker enhancers for internal breakers for VES-gelled aqueous fluids (Crews and Huang, 2010b). The oil-soluble surfactant breaker enhancers can overcome the rate-slowing effect that salinity has on the internal breakers, particularly at lower temperatures. In addition, oil-soluble surfactant breaker enhancers may allow lower internal breaker concentrations to be used to achieve quick and complete VES-gelled fluid breaking.

Oil-soluble surfactant breaker enhancers include various sorbitan (unsaturated) fatty acid esters (Crews and Huang, 2010a,b). These esters are mixed with mineral oils. Unsaturated fatty acids have been found to break down by autooxidation into VES breaking products or compositions. Each oil with various monoenoic and polyenoic acids uniquely shows the breakdown of the VES surfactant micelle structure by the presence of these autooxidation generated byproducts.

Various hydroperoxides may be formed in the course of these autooxidation reactions. The end products of these reactions typically include carbonyl compounds, alcohols, acids, and hydrocarbons. The rate of autooxidation of various fatty acids is shown in Table 16.5.

TABLE 16.5 Relative Rate of Autooxidation of C_{18} Acids (Crews and Huang, 2010b)

Fatty Acid	Double Bonds	Relative Oxidation Rate
Stearic	0	1
Oleic	1	100
Linoleic	2	1200
Linolenic	3	2500

Surfactant Polymer Compositions

The stability of viscoelastic surfactant-based fluids can be enhanced by using monomeric viscoelastic surfactants together with an oligomeric or polymeric compound with a thermally stable backbone. On this structure, viscoelastic surfactant functional groups are pending (Horton et al., 2009).

A viscoelastic surfactant solution can be synthesized from *N*-dodecene-1-yl-*N*,*N*-bis(2-hydroxyethyl)-*N*-methylammonium chloride in aqueous ammonium chloride. The aqueous solution is purged with nitrogen to remove residual oxygen and then oligomerized with 2,2′-azo(bis-amidinopropane)-dihydrochloride as radical initiator. The oligomerization is shown schematically in Figure 16.4. In the same way, potassium octadeceneoate,

$$CH_2{=}CH{-}(CH_2)_{15}{-}COO^-K^+,$$

can be oligomerized. The resulting oligomer can be imagined as related to an oligo ethylene backbone, to which relatively long pendant surfactant moieties are linked. If the double bonds are conjugated, as in the case of potassium octadecadieneoate,

$$CH_2{=}CH{-}CH{=}CH{-}(CH_2)_{13}COOK^+$$

the resulting oligomeric structure is related rather to poly(butadiene).

The oligomerization of the surfactant monomers in micelles effects that the viscosity of the gel becomes comparatively insensitive in contact with hydrocarbons. Further, the viscosity of the surfactant gel is hardly altered by the oligomerization of the surfactant monomers.

The preparation of several other oligomeric surfactants has been described in detail including the use of comonomers (Horton et al., 2007, 2009). An example for a cooligomer is shown in Figure 16.5.

The vicinal diol functionality renders the oligomers readily crosslinkable with polyvalent metal ions or complexes. The formulations have been characterized with respect to fluid loss control by an API standard (API Standard RP 13B-1, 2009).

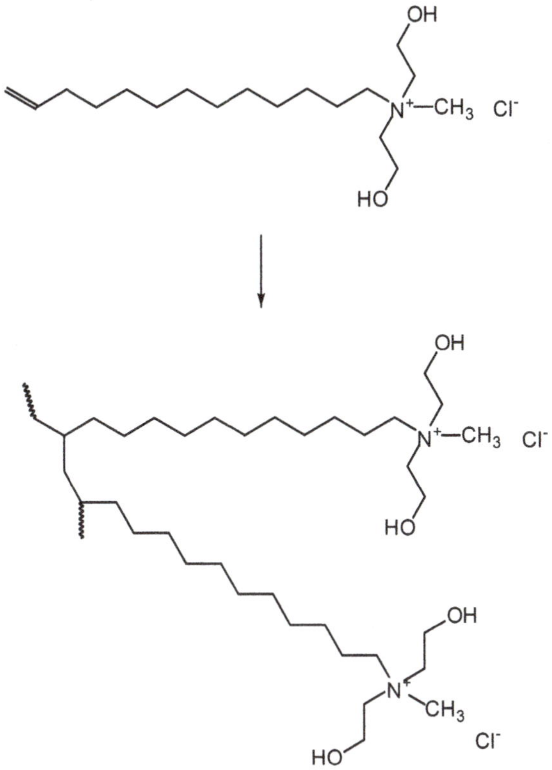

FIGURE 16.4 Oligomerization of unsaturated quaternary compounds (Horton et al., 2009).

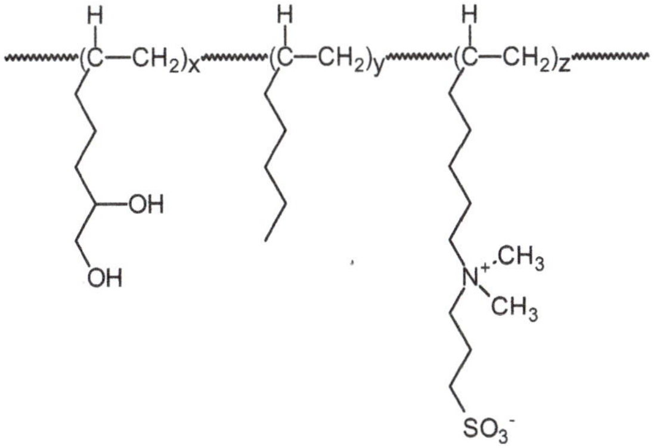

FIGURE 16.5 Cooligomeric surfactant (Horton et al., 2007).

Tradenames in References

Tradename	Supplier
Description	
Benol®	Sonneborn Refined Products
White mineral oil (Crews and Huang, 2010a,b)	
Captivates® liquid	ISP Hallcrest
Fish gelatin and gum acacia encapsulation coating (Crews and Huang, 2010a,b)	
Carnation®	Sonneborn Refined Products
White mineral oil (Crews and Huang, 2010a,b)	
ClearFRAC™	Schlumberger Technology Corp.
Stimulating Fluid (Crews, 2006,2010; Crews and Huang, 2010a,b)	
Diamond FRAQ™	Baker Oil Tools
VES breaker (Crews and Huang, 2010b)	
DiamondFRAQ™	Baker Oil Tools
VES System (Crews, 2010; Crews and Huang, 2010a,b)	
Escaid® (Series)	Crompton Corp.
Mineral oils (Crews and Huang, 2010a,b)	
Hydrobrite® 200	Sonneborn Inc.
White mineral oil (Crews and Huang, 2010a,b)	
Isopar® (Series)	Exxon
Isoparaffinic solvent (Crews and Huang, 2010a,b)	
Jordapon® ACI	BASF
Sodium cocoyl isothionate surfactant (Crews, 2010)	
Jordapon® CI	BASF
Ammonium cocoyl isothionate surfactant (Crews, 2010)	
Microsponge™	Advanced Polymer Systems
Porous solid substrate (Crews, 2006,2010; Crews and Huang, 2010a,b)	
Poly-S.RTM	Scotts Comp.
Polymer encapsulation coating (Crews, 2006,2010; Crews and Huang, 2010a,b)	
Span® 20	Uniqema
Sorbitan monolaurate (Crews and Huang, 2010b)	
Span® 40	Uniqema
Sorbitan monopalmitate (Crews and Huang, 2010b)	
Span® 61	Uniqema
Sorbitan monostearate (Crews and Huang, 2010b)	
Span® 65	Uniqema
Sorbitan tristearate (Crews and Huang, 2010b)	
Span® 80	Uniqema
Sorbitan monooleate (Crews and Huang, 2010b)	

(continued)

Continued

Tradename	Supplier
Description	
Span® 85	Uniqema
Sorbitan trioleate (Crews and Huang, 2010b)	
Tween® 20	Uniqema
Sorbitan monolaurate (Crews and Huang, 2010b)	
Tween® 21	Uniqema
Sorbitan monolaurate (Crews and Huang, 2010b)	
Tween® 40	Uniqema
Sorbitan monopalmitate (Crews and Huang, 2010b)	
Tween® 60	Uniqema
Sorbitan monostearate (Crews and Huang, 2010b)	
Tween® 61	Uniqema
Sorbitan monostearate (Crews and Huang, 2010b)	
Tween® 65	Uniqema
Sorbitan tristearate (Crews and Huang, 2010b)	
Tween® 81	Uniqema
Sorbitan monooleate (Crews and Huang, 2010b)	
Tween® 85	Uniqema
Sorbitan monooleate (Crews and Huang, 2010b)	
VES-STA 1	Baker Oil Tools
Gel stabilizer (Crews and Huang, 2010a,b)	
Wellguard™ 7137	Albemarle Corp.
Interhalogen gel breaker (Carpenter, 2007,2009)	
WG-3L VES-AROMOX® APA-T	Akzo Nobel
Viscoelastic surfactant (Crews and Huang, 2010a)	

REFERENCES

Acker, D.B., Malekahmadi, F., 2001. Delayed release breakers in gelled hydrocarbons. US Patent 6 187 720, 13 February 2001.

Ahlgren, J.A., 1993. Enzymatic hydrolysis of xanthan gum at elevated temperatures and salt concentrations. In: Proceedings Volume. Sixth Institute of Gas Technology Gas, Oil, & Environmental Biotechnology International Symposium, Colorado Springs, CO., 29 November–1 December 1993.

API Standard RP 13B-1, 2009. Recommended practice for field testing water-based drilling fluids. API Standard API RP 13B-1, American Petroleum Institute, Washington, DC.

Armstrong, C.D., 2012. Compositions useful for the hydrolysis of guar in high pH environments and methods related thereto. US Patent Application 20120111568, 10 May 2012. <http://www.freepatentsonline.com/20120111568.html>.

Barati, R., Johnson, S.J., McCool, S., Green, D.W., Willhite, G.P., Liang, J.-T., 2011. Fracturing fluid cleanup by controlled release of enzymes from polyelectrolyte complex nanoparticles. J. Appl. Polym. Sci. 121 (3), 1292–1298. http://dx.doi.org/10.1002/app.33343.

Barati, R., Johnson, S.J., McCool, S., Green, D.W., Willhite, G.P., Liang, J.-T., 2012. Polyelectrolyte complex nanoparticles for protection and delayed release of enzymes in alkaline pH and at elevated temperature during hydraulic fracturing of oil wells. J. Appl. Polym. Sci. 126 (2), 587–592. http://dx.doi.org/10.1002/app.36845.

Bielewicz, V.D., Kraj, L., 1998. Laboratory data on the effectivity of chemical breakers in mud and filtercake (Untersuchungen zur Effektivität von Degradationsmitteln in Spülungen). Erdöl Erdgas Kohle 114 (2), 76–79.

Boles, J.L., Metcalf, A.S., Dawson, J.C., 1996. Coated breaker for crosslinked acid. US Patent 5 497 830, assigned to BJ Services Co., 12 March 1996.

Brannon, H.D., Tjon-Joe-Pin, R.M., 1994. Biotechnological breakthrough improves performance of moderate to high-temperature fracturing applications. In: Proceedings Volume. 69th Annual SPE Technical Conference, vol. 1. New Orleans, 25–28 September 1994, pp. 515–530.

Cantu, L.A., Boyd, P.A., 1989. Laboratory and field evaluation of a combined fluid-loss control additive and gel breaker for fracturing fluids. In: Proceedings Volume. SPE Oilfield Chemistry International Symposium, Houston, 8–10 February 1989, pp. 7–16.

Cantu, L.A., McBride, E.F., Osborne, M., 1990a. Formation fracturing process. EP Patent 401 431, assigned to Conoco Inc. and Du Pont De Nemours & Co., 12 December 1990.

Cantu, L.A., McBride, E.F., Osborne, M., 1990b. Well treatment process. EP Patent 404 489, assigned to Conoco Inc. and Du Pont De Nemours & Co., 27 December 1990.

Carpenter, J.F., 2007. Breaker composition and process. US Patent 7 223 719, assigned to Albemarle Corporation, Richmond, VA, 29 May 2007. <http://www.freepatentsonline.com/7223719.html>.

Carpenter, J.F., 2009. Bromine-based sulfamate stabilized breaker composition and process. US Patent 7 576 041, assigned to Albemarle Corporation, Baton Rouge, LA, 18 August 2009. <http://www.freepatentsonline.com/7576041.html>.

Cordova, M., Cheng, M., Trejo, J., Johnson, S.J., Willhite, G.P., Liang, J.-T., Berkland, C., 2008. Delayed HPAM gelation via transient sequestration of chromium in polyelectrolyte complex nanoparticles. Macromolecules 41 (12), 4398–4404. http://dx.doi.org/10.1021/ma800211d.

Craig, D., 1991. The degradation of hydroxypropyl guar fracturing fluids by enzyme, oxidative, and catalyzed oxidative breakers. Ph.D. Thesis, Texas A & M University.

Craig, D., Holditch, S.A., 1993a. The degradation of hydroxypropyl guar fracturing fluids by enzyme, oxidative, and catalyzed oxidative breakers: Pt.1: linear hydroxypropyl guar solutions: topical report, February 1991–December 1991. Gas. Res. Inst. Rep. GRI-93/04191, Gas. Res. Inst.

Craig, D., Holditch, S.A., 1993b. The degradation of hydroxypropyl guar fracturing fluids by enzyme, oxidative, and catalyzed oxidative breakers: Pt.2: crosslinked hydroxypropyl guar gels: topical report, January 1992–April 1992. Gas. Res. Inst. Rep. GRI-93/04192, Gas. Res. Inst.

Craig, D., Holditch, S.A., Howard, B., 1992. The degradation of hydroxypropyl guar fracturing fluids by enzyme, oxidative, and catalyzed oxidative breakers. In: Proceedings Volume. 39th Annual Southwestern Petroleum Short Course Association Inc. et. al Meeting, Lubbock, TX, 22–23 April 1992, pp. 1–19.

Crews, J.B., 2006. Bacteria-based and enzyme-based mechanisms and products for viscosity reduction breaking of viscoelastic fluids. US Patent 7 052 901, assigned to Baker Hughes Incorporated, Houston, TX, 30 May 2006. <http://www.freepatentsonline.com/7052901.html>.

Crews, J.B., 2007a. Aminocarboxylic acid breaker compositions for fracturing fluids. US Patent 7 208 529, assigned to Baker Hughes Incorporated Houston, TX, 24 April 2007. <http://www.freepatentsonline.com/7208529.html>.

Crews, J.B., 2007b. Polyols for breaking of fracturing fluid. US Patent 7 160 842, assigned to Baker Hughes Incorporated, Houston, TX, 9 January 2007. <http://www.freepatentsonline.com/7160842.html>.

Crews, J.B., 2010. Saponified fatty acids as breakers for viscoelastic surfactant-gelled fluids. US Patent 7 728 044, assigned to Baker Hughes Incorporated, Houston, TX, 1 June 2010. <http://www.freepatentsonline.com/7728044.html>.

Crews, J.B., Huang, T., 2010a. Unsaturated fatty acids and mineral oils as internal breakers for ves-gelled fluids. US Patent 7 696 134, assigned to Baker Hughes Incorporated, Houston, TX, 13 April 2010. <http://www.freepatentsonline.com/7696134.html>.

Crews, J.B., Huang, T., 2010b. Use of oil-soluble surfactants as breaker enhancers for ves-gelled fluids. US Patent 7 696 135, assigned to Baker Hughes Incorporated, Houston, TX, 13 April 2010. <http://www.freepatentsonline.com/7696135.html>.

Dawson, J.C., Le, H.V., 1995. Controlled degradation of polymer based aqueous gels. US Patent 5 447 199, assigned to BJ Services Co., 5 September 1995.

Fodge, D.W., Anderson, D.M., Pettey, T.M., 1996. Hemicellulase active at extremes of pH and temperature and utilizing the enzyme in oil wells. US Patent 5 551 515, assigned to Chemgen Corp., 3 September 1996.

Gulbis, J., King, M.T., Hawkins, G.W., Brannon, H.D., 1990a. Encapsulated breaker for aqueous polymeric fluids. In: Proceedings Volume. 9th SPE Formation Damage Control Symposium, Lafayette, LA, 22–23 February 1990, pp. 245–254.

Gulbis, J., Williamson, T.D.A., King, M.T., Constien, V.G., 1990b. Method of controlling release of encapsulated breakers. EP Patent 404 211, assigned to Pumptech NV and Dowell Schlumberger SA, 27 December 1990.

Gulbis, J., King, M.T., Hawkins, G.W., Brannon, H.D., 1992. Encapsulated breaker for aqueous polymeric fluids. SPE Prod. Eng. 7 (1), 9–14.

Gupta, D.V.S., Cooney, A., 1992. Encapsulations for treating subterranean formations and methods for the use thereof. WO Patent 9210640, assigned to Western Co., North America, 25 June 1992.

Gupta, D.V.S., Prasek, B.B., 1995. Method for fracturing subterranean formations using controlled release breakers and compositions useful therein. US Patent 5 437 331, assigned to Western Co., North America, 1 August 1995.

Harms, W.M., 1992. Catalyst for breaker system for high viscosity fluids. US Patent 5 143 157, assigned to Halliburton Co., 1 September 1992.

Harris, R.E., Hodgson, R.J., 1998. Delayed acid for gel breaking. US Patent 5 813 466, assigned to Cleansorb Limited, GB, 29 September 1998. <http://www.freepatentsonline.com/5813466.html>.

Hawkins, G.W., 1986. Molecular weight reduction and physical consequences of chemical degradation of hydroxypropylguar in aqueous brine solutions. In: Proceedings 192nd ACS National Meetings American Chemical Society Polymeric Materials Science Engineering Division of Technology Program, vol. 55. Anaheim, Calif, 7–12 September 1986, pp. 588–593.

Horton, R.L., Prasek, B., Growcock, F.B., Kippie, D., Vian, J.W., Abdur-Rahman, K.B., Arvie, Jr., M., 2007. Surfactant-polymer compositions for enhancing the stability of viscoelastic-surfactant based fluid. US Patent 7 157 409, assigned to M-I LLC, Houston, TX, 2 January 2007. <http://www.freepatentsonline.com/7157409.html>.

Horton, R.L., Prasek, B.B., Growcock, F.B., Kippie, D.P., Vian, J.W., Abdur-Rahman, K.B., Arvie, M., 2009. Surfactant-polymer compositions for enhancing the stability of viscoelastic-surfactant based fluid. US Patent 7 517 835, assigned to M-I LLC, Houston, TX, April 14 2009. <http://www.freepatentsonline.com/7517835.html>.

Hunt, C.V., Powell, R.J., Carter, M.L., Pelley, S.D., Norman, L.R., 1997. Encapsulated enzyme breaker and method for use in treating subterranean formations. US Patent 5 604 186, assigned to Halliburton Co., 18 February 1997.

Jihua, C., Sui, G., 2011. Rheological behaviors of bio-degradable drilling fluids in horizontal drilling of unconsolidated coal seams. Int. J. Info. Technol. Comput. Sci. 3 (3), 1–7.

Johnson, S., Barati, R., McCool, S., Green, D.W., Willhite, G.P., Liang, J.-T., 2011. Polyelectrolyte complex nanoparticles to entrap enzymes for hydraulic fracturing fluid cleanup. Preprints Amer. Chem. Soc. Div. Pet. Chem. 56 (2), 168–172. <http://pubs.acs.org/cgi-bin/preprints/display?div=petr&meet=242&page=88718.pdf>.

King, M.T., Gulbis, J., Hawkins, G.W., Brannon, H.D., 1990. Encapsulated breaker for aqueous polymeric fluids. In: Proceedings Volume. Canadian Institute of Mining, Metallurgy and Petroleum Society/SPE International Technology Meeting, vol. 2. Calgary, Canada, 10–13 June 1990.

Kyaw, A., Binti, Nor Azahar, B.S., Tunio, S.Q., 2012. Fracturing fluid (guar polymer gel) degradation study by using oxidative and enzyme breaker. Res. J. Appl. Sci. Eng. Technol. 4 (12), 1667–1671. <http://maxwellsci.com/print/rjaset/v4-1667-1671.pdf>.

Langemeier, P.W., Phelps, M.A., Morgan, M.E., 1989. Method for reducing the viscosity of aqueous fluids. EP Patent 330489, 30 August 1989.

Laramay, S.B., Powell, R.J., Pelley, S.D., 1995. Perphosphate viscosity breakers in well fracture fluids. US Patent 5 386 874, assigned to Halliburton Co., 7 February 1995.
Ma, Y., Xue, Y., Dou, Y., Xu, Z., Tao, W., Zhou, P., 2004. Characterization and gene cloning of a novel/?-mannanase from alkaliphilic bacillus sp. n16–5. Extremophiles 8 (6), 447–454. http://dx.doi.org/10.1007/s00792-004-0405-4.
Manalastas, P.V., Drake, E.N., Kresge, E.N., Thaler, W.A., McDougall, L.A., Newlove, J.C., Swarup, V., Geiger, A.J., 1992. Breaker chemical encapsulated with a crosslinked elastomer coating. US Patent 5 110486, assigned to Exxon Research & Eng. Co., 5 May 1992.
McDougall, L.A., Malekahmadi, F., Williams, D.A., 1993. Method of fracturing formations. EP Patent 540 204, assigned to Exxon Chemical Patents In, 5 May 1993.
Mitchell, T.O., Card, R.J., Gomtsyan, A., 2001. Cleanup additive. US Patent 6 242 390, assigned to Schlumberger Technol. Corp., 5 June 2001.
Mondshine, T.C., 1993. Process for decomposing polysaccharides in alkaline aqueous systems. US Patent 5 253 711, assigned to Texas United Chemical Corp., Houston, TX, 19 October 1993. <http://www.freepatentsonline.com/5253711.html>.
Muir, D.J., Irwin, M.J., 1999. Encapsulated breakers, compositions and methods of use. WO Patent 9 961 747, assigned to 3M Innovative Propertie C, 2 December 1999.
Noran, L., Vitthal, S., Terracina, J., 1995. New breaker technology for fracturing high-permeability formations. In: Proceedings Volume. SPE Europe Formation Damage Control Conference, The Hague, Neth, 15–16 May 1995, pp. 187–199.
Norman, L.R., Laramay, S.B., 1994. Encapsulated breakers and method for use in treating subterranean formations. US Patent 5 373 901, assigned to Halliburton Co., 20 December 1994.
Norman, L.R., Turton, R., Bhatia, A.L., 2001. Breaking fracturing fluid in subterranean formation. EP Patent 1 152 121, assigned to Halliburton Energy Serv., 7 November 2001.
Prasek, B.B., 1996. Interactions between fracturing fluid additives and currently used enzyme breakers. In: Proceedings Volume. 43rd Annual Southwestern Petroleum Short Course Association Inc. et al. Meeting Lubbock, Texas, 17–18 April 1996, pp. 265–279.
Sarwar, M.U., Cawiezel, K., Nasr-El-Din, H., 2011. Gel degradation studies of oxidative and enzyme breakers to optimize breaker type and concentration for effective break profiles at low and medium temperature ranges. In: Proceedings of SPE Hydraulic Fracturing Technology Conference. No. 140520-MS. SPE Hydraulic Fracturing Technology Conference, 24–26 January 2011, The Woodlands, Texas, USA, Society of Petroleum Engineers, Dallas, Texas, pp. 1–21. http://dx.doi.org/10.2118/140520-MS.
Shuchart, C.E., Terracina, J.M., Slabaugh, B.F., McCabe, M.A., 1999. Method of treating subterranean formation. EP Patent 916 806, assigned to Halliburton Energy Serv., 19 May 1999.
Slodki, M.E., Cadmus, M.C., 1991. High-temperature, salt-tolerant enzymic breaker of xanthan gum viscosity. In: Donaldson, E.C. (Ed.), Microbial Enhancement of Oil Recovery: Recent Advances: Proceedings of the 1990 International Conference on Microbial Enhancement of Oil Recovery of Developments in Petroleum Science, vol. 31. Elsevier Science Ltd., pp. 247–255.
Swarup, V., Peiffer, D.G., Gorbaty, M.L., 1996. Encapsulated breaker chemical. US Patent 5 580 844, assigned to Exxon Research & Eng. Co., 3 December 1996.
Syrinek, A.R., Lyon, L.B., 1989. Low temperature breakers for gelled fracturing fluids. US Patent 4 795 574, assigned to Nalco Chemical Co., 3 January 1989.
Tjon-Joe-Pin, R.M., 1993. Enzyme breaker for galactomannan based fracturing fluid. US Patent 5 201 370, assigned to BJ Services Company, Houston, TX, 13 April 1993. <http://www.freepatentsonline.com/5201370.html>.
Walker, M.L., Shuchart, C.E., 1995. Method for breaking stabilized viscosified fluids. US Patent 5 413 178, assigned to Halliburton Co., 9 May 1995.
Williams, M.M., Phelps, M.A., Zody, G.M., 1987. Reduction of viscosity of aqueous fluids. EP Patent 222 615, 20 May 1987.

Chapter 17

Biocides

The base carrier fluids for fracturing treatments are most often shallow water wells, streams, or ponds. This water is often laden with bacterial growth and saturated with dissolved oxygen requiring the fluids be treated with biocides and oxygen scavengers during the fracture stimulation to prevent accelerated corrosion of the equipment.

Unfortunately, the most commonly used biocides and oxygen scavengers either negatively react with one another or with other compounds in the fracturing fluid. The interactions and effects of various biocides and oxygen scavengers with components in fracturing fluids have been detailed (McCurdy, 2008).

Major problems in oil and gas operations result from the biogenic formation of hydrogen sulfide (H_2S) in the reservoir. The presence of H_2S results in increased corrosion, iron sulfide formation, higher operating costs, and reduced revenue and constitutes a serious environmental and health hazard.

Severe corrosion also can result from the production of acids associated with the growth of certain bacterial biofilms. These bacterial biofilms are often composed of sulfate-reducing bacteria, which grow anaerobically in water, often in the presence of oil and natural gases. Once biofilms are established, it is extremely difficult to regain the biologic control of the system.

When biofilms are formed on metallic surfaces, they can seriously corrode oil production facilities. Microbiologically influenced corrosion represents the most serious form of that degradation.

It is estimated that microbiologically influenced corrosion may be responsible for 15–30% of failures caused by corrosion in all industries.

Thus an effective control of bacteria, which are responsible for these undesired effects, is mandatory. Several biocides and nonbiocidal techniques to control bacterial corrosion are available. Procedures and techniques to detect bacteria have been developed.

Hydraulic Fracturing Chemicals and Fluids Technology. http://dx.doi.org/10.1016/B978-0-12-411491-3.00017-0
© 2013 Elsevier Inc. All rights reserved.

MECHANISMS OF GROWTH

Growth of Bacteria Supported by Oilfield Chemicals

Growth experiments have been conducted using bacteria from oil installations with several chemicals normally used in injection water treatments. The studies revealed that some chemicals could be utilized as nitrogen sources, as phosphorus sources, and as carbon sources for those bacteria (Sunde et al., 1990). Therefore, it is concluded that the growth potential of water treatment additives may be substantial and should be investigated before the selection of the respective chemicals.

In other experiments it was established that the cultures of sulfate-reducing bacteria isolated from waters of several oilfields have a greater capacity to form H_2S than the collection culture of sulfate-reducing bacteria used as the standard in these investigations. The stimulating effect of the chemical products on individual elective cultures of sulfate-reducing bacteria can vary considerably depending on the species, activity, and adaptation of bacteria to the chemical products.

Selective cultures of sulfate-reducing bacteria as a result of adaptation acquire relative resistance to toxic compounds. Thus, when bactericides are used for complete suppression of the vital activity of sulfate-reducing bacteria in the bottomhole zone and reservoir, higher doses of the bactericide than those calculated for laboratory collection cultures are necessary (Kriel et al., 1993).

It has been shown that sulfidogenic bacteria injected into a reservoir with floodwater may survive higher temperatures in the formation and can be recovered from producing well fluids (Salanitro et al., 1993). These organisms may colonize cooler zones and sustain growth by degrading fatty acids in formation waters.

Mathematical Models

A mathematical model for reservoir souring caused by the growth of sulfate-reducing bacteria is available. The model is a one-dimensional numerical transport model based on conservation equations and includes bacterial growth rates and the effect of nutrients, water mixing, transport, and adsorption of H_2S in the reservoir formation. The adsorption of H_2S by the rock has been considered. Two basic concepts for microbial H_2S production were tested with field data (Sunde et al., 1993):

- H_2S production in the mixing zone between formation water and injection water (mixing zone model) and
- H_2S production caused by the growth of sulfate-reducing bacteria in a biofilm in the reservoir rock close to the injection well (biofilm model).

Field data obtained from three oil producing wells on the Gullfaks field correlated with H_2S production profiles obtained using the biofilm model but could not be explained by the mixing zone model (Sunde et al., 1993).

Model of Colony Growth

The growth of bacteria with time in the presence of various amounts of copper sulfate is shown in Figure 17.1. As a measure for the growth serves the diameter of the colonies.

A simplified model of colony growth has been presented (Rodin et al., 2005). According to this model, during growth, a colony passes successively through exponential and linear phases of growth. However, the exponential phase of growth persists unless the concentration of a nutritious substrate becomes limited. The increase of colony diameter d during the exponential phase can be described as

$$d = d_0 \exp\left(\frac{\mu'_m}{2} t\right). \tag{17.1}$$

Here d is the diameter of the colony at the incubation time t, d_0 is the effective diameter of the individual cell, the colony progenitor, and μ'_m is the maximal growth rate.

In contrast, the linear phase of growth occurs at nutrient limitation and it starts at a time t_l, after which the colony diameter increases with constant rate

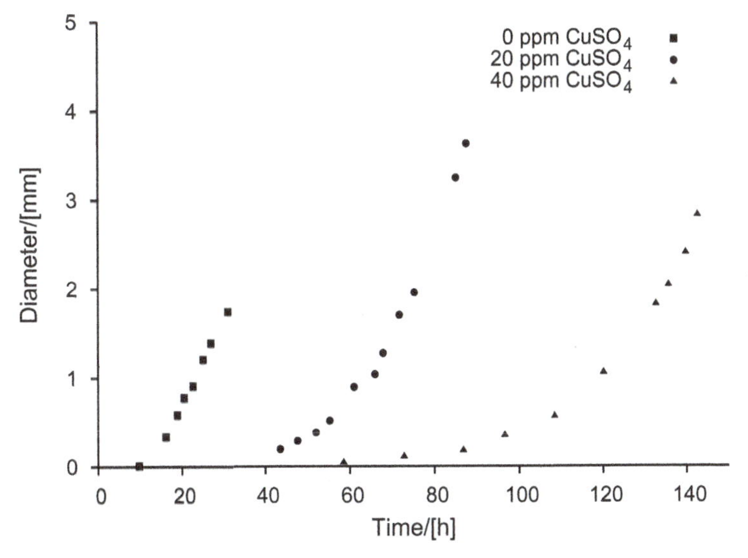

FIGURE 17.1 Effect of copper sulfate on the growth of *S. marcescens* colonies (Rodin et al., 2005).

k_d according to

$$d = d_{t_l} + k_d(t - t_l). \tag{17.2}$$

Equations (17.1) and (17.2) can be combined to give

$$t = \frac{d}{k_d} + \frac{2}{\mu'_m}\left[\ln\left(\frac{2k_d}{d_0\,\mu'_m}\right) - 1\right]. \tag{17.3}$$

Equation (17.3) relates the key parameter of the exponential growth phase of an invisible microcolony with the parameters of the linear growth phase of the visible macrocolony. This property essentially simplifies the experimental determination of the parameter μ'_m (Rodin et al., 2005).

Sulfate-Reducing Bacteria

Sulfate-reducing bacteria belong to the class of chemolithotrophic bacteria (Barton and Fauque, 2009). Around 220 species of 60 genera are known. All use sulfate as terminal electron acceptor. In this way, they constitute a unique physiological group of microorganisms that couple anaerobic electron transport to ATP synthesis. New strains of sulfate-reducing bacteria are continuously discovered (Miranda-Tello et al., 2003; Youssef et al., 2009; Agrawal et al., 2010). For example in 2003, a new spirilloid sulfate-reducing bacterium designated strain MET2T was isolated from a Mexican oilfield separator (Miranda-Tello et al., 2003).

These bacteria can use a wide variety of compounds as electron donors. Proteins with metal groups that can be oxidized and reduced are the basis of metabolism. In particular, they act on soluble electron transfer proteins and via transmembrane redox complexes. The capability of sulfate-reducing bacteria to utilize hydrocarbons in pure cultures and consortia offers the possibility for the bioremediation of soils that are contaminated with aromatic hydrocarbons.

Some strains of sulfate-reducing bacteria can even reduce chlorinated compounds, e.g., 3-chlorobenzoate, chloroethenes, and nitroaromatic compounds. Sulfate-reducing bacteria can reduce also some heavy metals. Therefore, several procedures have been proposed for using these strains in the bioremediation of toxic metals.

In the course of metabolism, high levels of hydrogen sulfide are produced. Since the produced sulfide contributes to the souring of the oilfields, corrosion of casings and concrete is a serious problem (Barton and Fauque, 2009).

Bacterial Corrosion

Bacterial corrosion is often referred to as *microbiologically influenced corrosion*. Microbiologically influenced corrosion involves the initiation or acceleration of corrosion by microorganisms. The metabolic products of microorganisms appear to affect most engineering materials, but the more commonly used corrosion resistant alloys, such as stainless steels, seem to be particularly susceptible.

The importance of microbiologically influenced corrosion has been underestimated, because most microbiologically influenced corrosion occurs as a localized, pitting-type attack. In general, this corrosion type results in relatively low rates of weight loss, changes in electrical resistance, and changes in total area affected. This makes microbiologically influenced corrosion difficult to detect and to quantify using traditional methods of corrosion monitoring (Pope et al., 1992).

pH Regulation

Weak acids are products of bacterial metabolisms. Sulfate-reducing bacteria regulate the pH of their environment at levels that depend on the potential secondary reactions:

- Precipitation of iron sulfide,
- Oxidation of sulfide ions to thiosulfate by traces of oxygen, and
- Metabolism of this thiosulfate or of other sulfur compounds.

In this way, it is possible to explain the initiation and growth of bacterial corrosion pits.

Biocide Enhancers

In order to effectively treat water against bacterial contamination, a fast-acting biocide is needed. This may be even more important for on-the-fly treatments, where biocides have a very short contact time with the water before other treating chemicals are added and the fluids are pumped downhole. To get an effective fast kill of bacteria, in some instances it is believed to be helpful to include a biocide enhancer to aid the biocide treatments or work synergistically with the biocide (Bryant et al., 2009).

Quaternary surfactants may act at the same time as biocide enhancers. For example, 19N™ is a cationic surfactant that also is a biocide enhancer. When used in combination with biocides such as sodium hypochlorite or glutaraldehyde, in some instances, bacterial problems may be treated in short times in the range of 5 min.

Although quaternary surfactants can be useful when used together with biocides, some quaternary surfactants may have a fundamental incompatibility with anionic friction reducers, which are also useful in subterranean operations.

It is believed that this fundamental incompatibility may arise from charges present on both molecules that may cause the quaternary surfactant and the friction reducer to react and eventually a precipitate is formed. Additionally, some biocides, e.g., such as oxidizers, may degrade certain friction reducers (Bryant et al., 2009).

PERFORMANCE CONTROL

Bacterial contamination of fracturing fluids may cause a number of serious problems in the oilfield. If the fracturing fluids are poorly treated or untreated, sulfate-reducing bacteria and acid-producing bacteria can become established.

Fracturing fluids contain often poly(acrylamide) or sugar-based polymers and other organic compounds which are potential food sources for bacteria.

Usually, the water used for preparation of these fluids is obtained from rivers, lakes, or oilfield wastewater which may be heavily contaminated with bacteria.

Various biocides were tested, including bronopol, glutaraldehyde, a glutaraldehyde/quaternary ammonium compound blend, isothiazolin, tetrakishydromethyl phosphonium sulfate, and 2,2-dibromo-3-nitrilopropionamide. Tetrakishydromethyl phosphonium sulfate performed well providing both quick kill of aerobic, fermentative, and sulfate-reducing bacteria. Also a long-term preservation of the produced fluid was observed. In contrast, glutaraldehyde as such and glutaraldehyde blends did not perform well on the general aerobic and acid-producing bacteria (Johnson et al., 2008; Fichter et al., 2009).

The biocides were tested in fluids used for fracturing Barnett shales in the Fort Worth basin of Texas. The Barnett shale is a one and a half mile deep underground natural gas formation that extends over 5000 square miles. This field is the second largest oil and gas field in the United States (Johnson et al., 2008).

TREATMENTS WITH BIOCIDES

Previously Fractured Formations

A special problem is the refracturing of a previously fractured formation that is contaminated with bacteria. In such a case, the fracturing fluid must be mixed with an amount of biocide sufficient to reach and to kill the bacteria contained in the formation. The refracturing of the formation causes the bactericide to be distributed throughout the formation and to contact and kill bacteria contained therein (McCabe et al., 1991).

Intermittent Addition of Biocide

The intermittent addition technique is as (Hegarty and Levy (1996a,b,c)):

- The addition of a slug dose of a biologically effective amount of a quick kill biocide.
- Intermittent addition of biologically effective amounts of a control biocide. This means that the control biocide is dosed for a certain period of time, followed by a period of much lower or zero dosing. This cycle is repeated throughout the treatment.

This process reduces the amount of control biocide employed in the control of contamination of oil production system waters by sessile bacteria. The biocide may be applied at intervals of 2–15 d. The duration of biocide application is preferably from 4 to 8 h (Moody and Montgomerie, 1996).

Nonbiocidal Control

Chemical treatments for bacteria control represent significant cost and environmental liability. Because the regulatory pressure on the use of toxic biocides is increasing, more environmentally acceptable control measures are being developed.

Biocompetitive Exclusion Technology

Besides adding biocides to wells, another approach seems to be promising in modifying the reservoir ecology. The production of sulfide can be decreased, and its concentration is reduced by the establishment and growth of an indigenous microbial population that replaces the population of sulfate-reducing bacteria.

The technology is based on the addition of low concentrations of a water-soluble nutrient solution that selectively stimulates the growth of an indigenous microbial population, thereby inhibiting the detrimental sulfate-reducing bacteria population that causes the generation of H_2S. This deliberate and controlled modification of the microflora and reservoir ecology has been termed *biocompetitive exclusion* (Sandbeck and Hitzman, 1995; Hitzman and Dennis, 1997).

Inhibitors for Bacterial Films

Laboratory tests with quaternary amine additives showed a very low surface colonization and lower corrosion rates (Enzien et al., 1996). On the other hand, the biocidal effect of quaternary amines in the test fluids appeared to be minimal. These results suggest that quaternary amines may prevent microbiologically influenced corrosion by mechanisms other than killing bacteria and that treatments preventing colonization on the surface may persist longer than most biocides.

Periodic Change in Ionic Strengths

For the effective control of microorganisms, it is necessary to take into account the mechanism of formation of bacteria and the ecologic factors affecting it. The process of vital activity of bacteria begins with their adsorption on the enclosing rocks and adaptation to the new habitat conditions. Pure cultures of sulfate-reducing bacteria are not active in crude oil.

Their development in oil reservoirs depends entirely on hydrocarbon-oxidizing bacteria, which are the primary cause of breakdown of oil. If the ecologic conditions in the reservoir are changed during formation of the microorganisms, the established food chains are disrupted and the active development of microflora ceases. It was experimentally established that the periodic injection of waters markedly differing in mineralization, taking into account the ecologic characteristics of the formation of the microorganisms, makes it possible to control biogenic processes in an oil reservoir without disturbing the sanitary state of the environment (Blagov et al., 1990).

SPECIAL CHEMICALS

Various biocide technologies have been used successfully in water treatment applications for many years. These include oxidizers, such as chlorine and bromine products, and nonoxidizing biocides, including isothiazolones, quats, organobromines, and glutaraldehyde.

The efficiency of multiple biocides that are commonly used to control sulfate-reducing bacteria in fracturing fluids in shale natural gas formations has been critically evaluated (Struchtemeyer et al., 2012). Biocides, including tetrakis(hydroxymethyl) phosphonium sulfate, sodium hypochlorite, didecyldimethylammonium chloride, tri-*n*-butyl tetradecyl phosphonium chloride, glutaraldehyde, were tested. The minimum inhibitory concentrations were determined with planktonic cells and biofilms. The results indicated that the biofilm formation by sulfate-reducing bacteria negatively influences the efficiency of the biocides.

Biocides are often misapplied in the petroleum industry. Many of the misapplications occur because the characteristics of the biocides are not considered before use. Some guidelines of biocide selection are outlined in a review in the literature (Boivin, 1994). The early detection of microbiologic problems is imperative, and reparative actions must be taken as soon as possible.

These measurements should include changes in operating methods to prevent degradation of the operating environment. This might include the rejection of untreated waters for cleaning deposits in vessels and lines. In general, biocides are needed to control the activity of the bacteria in a system. However, biocides alone usually will not solve a microbiologic problem. Five requirements for the bactericide selection are emphasized (Zhou, 1990):

1. Wide bacteria-killing ability and range.
2. Noncorrosive property, good inhibiting ability, and convenience for transportation and application.
3. Nontoxic or low toxicity property that causes no damage to human beings and is within environmental control regulations.
4. Good miscibility, with no damage or interference to drilling fluid or its chemical agents.
5. Bacteria-killing effect that is not affected by environmental adaptation of the bacteria.

A hydraulic fracturing fluid containing guar gum or other natural polymers can be stabilized against bacterial attack by adding heterocyclic sulfur compounds. This method of stabilization prevents any undesired degradation of the fracturing fluid, such as reduction of its rheological properties (which are necessary for conducting the hydraulic fracturing operation), at high

TABLE 17.1 Biocides (Kanda et al., 1988; Kanda and Kawamura, 1989)

Compound	Compound
Mercaptobenzimidazole[a]	1,3,4-Thiadiazole-2,5-dithiol[a,b]
2-Mercaptobenzothiazole	2-Mercaptothiazoline
2-Mercaptobenzoxazole	2-Mercaptothiazoline
2-Thioimidazolidone	2-Thioimidazoline
4-Ketothiazolidine-2-thiol	*N*-Pyridineoxide-2-thiol

[a]*For Guar gum.*
[b]*For Xanthan gum.*

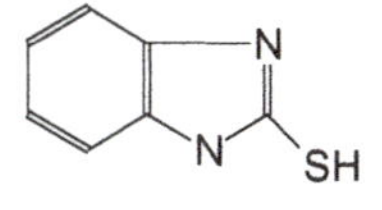

2-Mercaptobenzoimidazole

2-Mercaptobenzothiazole

2-Mercaptobenzoxazole

2-Mercaptothiazoline

2,5-Dimercapto-1,3,4-thiadiazole

2-Imidazolidinethione

4-Ketothiazolidine-2-thiol

Pyridine-N-oxide-2-thiol

FIGURE 17.2 Biocides for hydraulic fracturing fluids.

4-Ketothiazolidine-2-thiol Pyridine-N-oxide-2-thiol

FIGURE 17.3 Thiols.

temperatures. Biocides suitable for fracturing fluids are shown in Table 17.1 and Figures 17.2 and 17.3.

REFERENCES

Agrawal, A., Vanbroekhoven, K., Lal, B., 2010. Diversity of culturable sulfidogenic bacteria in two oil-water separation tanks in the north-eastern oil fields of india. Anaerobe 16 (1), 12–18. http://www.sciencedirect.com/science/article/B6W9T-4W7J150-1/2/bd2cb4d4c087315b0ed914986b4ed3c2.

Barton, L.L., Fauque, G.D., 2009. Biochemistry, physiology and biotechnology of sulfate-reducing bacteria. In: Laskin, A.I., Sariaslani, S., Gadd, G.M. (Eds.), Advances in Applied Microbiology, vol. 68. Academic Press, pp. 41–98 (Chapter 2). <http://www.sciencedirect.com/science/article/B7CSY-4W79HR5-4/2/1b6c4d955860bf7b7eecc73674593b03>.

Blagov, A.V., Prazdnikova, Z.F., Praporshchikov, V.I., 1990. Use of ecological factors for controlling biogenic sulfate reduction. Neft Khoz 5, 48–50.

Boivin, J., 1994. Oil industry biocides. Mater. Perf. 34 (2), 65–68.

Bryant, J.E., McMechan, D.E., McCabe, M.A., Wilson, J.M., King, K.L., 2009. Treatment fluids having biocide and friction reducing properties and associated methods. US Patent Application 20090229827, 17 September 2009. <http://www.freepatentsonline.com/20090229827.html>.

Enzien, M.V., Pope, D.H., Wu, M.M., Frank, J., 1996. Nonbiocidal control of microbiologically influenced corrosion using organic film-forming inhibitors. In: Proceedings Volume. 51st Annual NACE International Corrosion Conference (Corrosion 96) Denver, 24–29 March 1996.

Fichter, J.K., Johnson, K., French, K., Oden, R., 2009. Biocides control barnett shale fracturing fluid contamination. Oil Gas J. 107 (19), 38–44.

Hegarty, B.M., Levy, R., 1996a. Control of oilfield biofouling. CA Patent 2160305, 13 April 1996.

Hegarty, B.M., Levy, R., 1996b. Control of oilfield biofouling. EP Patent 706759, 17 April 1996.

Hegarty, B.M., Levy, R., 1996c. Procedure for combatting biological contamination in petroleum production (procede pour combattre l'encrassement biologique dans la production de petrole). FR Patent 2725754, 19 April 1996.

Hitzman, D.O., Dennis, D.M., 1997. Sulfide removal and prevention in gas wells. In: Proceedings Volume. SPE Production and Operations Symposium, Oklahoma City, 9–11 March 1997, pp. 433–438.

Johnson, K., French, K., Fichter, J.K., Oden, R., 2008. Use of microbiocides in Barnett shale gas well fracturing fluids to control bacteria related problems. In: Corrosion, 2008. NACE International, New Orleans, LA.

Kanda, S., Kawamura, Z., 1989. Stabilization of xanthan gum in aqueous solution. US Patent 4810786, 7 March 1989.

Kanda, S., Yanagita, M., Sekimoto, Y., 1988. Stabilized fracturing fluid and method of stabilizing fracturing fluid. US Patent 4721577, 26 January 1988.

Kriel, B.G., Crews, A.B., Burger, E.D., Vanderwende, E., Hitzman, D.O., 1993. The efficacy of formaldehyde for the control of biogenic sulfide production in porous media. In: Proceedings Volume. SPE Oilfield Chemistry International Symposium, New Orleans, 2–5 March 1993, pp. 441–448.

McCabe, M.A., Wilson, J.M., Weaver, J.D., Venditto, J.J., 1991. Biocidal well treatment method. US Patent 5016714, assigned to Halliburton Co., 21 May 1991.

McCurdy, R., 2008. Selecting and applying biocides and oxygen scavengers in high volume, high rate hydraulic fracture stimulations. Proc. Annu. SW Petrol. Short Course 55, 357–366.

Miranda-Tello, E., Fardeau, M.-L., Fernández, L., Ramírez, F., Cayol, J.-L., Thomas, P., Garcia, J.-L., Ollivier, B., 2003. Desulfovibrio capillatus sp. nov., a novel sulfate-reducing bacterium isolated from an oil field separator located in the gulf of mexico. Anaerobe 9 (2), 97–103. <http://www.sciencedirect.com/science/article/B6W9T-48Y6FFJ-5/2/8e1212c71cc036aa9b6bb5ef66e0e5e7>.

Moody, S.S., Montgomerie, H.T.R., 1996. Control of oilfield biofouling. EP Patent 706974, 17 April 1996.

Pope, D.H., Dziewulski, D.M., Lockwood, S.F., Werner, D.P., Frank, J.R., 1992. Microbiological corrosion concerns for pipelines and tanks. In: Proceedings Volume. API Pipeline Conference, Houston, 7–8 April 1992, pp. 290–321.

Rodin, V.B., Zhigletsova, S.K., Kobelev, V.S., Akimova, N.A., Kholodenko, V.P., 2005. Efficacy of individual biocides and synergistic combinations. Int. Biodeterioration Biodegrad. 55 (4), 253–259. <http://www.sciencedirect.com/science/article/B6VG6-4G4MMGS-1/2/3169cb82006ef93b1ab178e3816004a3>.

Salanitro, J.P., Williams, M.P., Langston, G.C., 1993. Growth and control of sulfidogenic bacteria in a laboratory model seawater flood thermal gradient. In: Proceedings Volume. SPE Oilfield Chemistry International Symposium, New Orleans, 2–5 March 1993, pp. 457–467.

Sandbeck, K.A., Hitzman, D.O., 1995. Biocompetitive exclusion technology: A field system to control reservoir souring and increase production. In: US DOE Rep. No. CONF-9509173. 5th US DOE et al Microbial Enhanced Oil Recovery & Related Biotechnology for Solving Environment Problems International Conference, Dallas, 11–14 September 1995, pp. 311–319.

Struchtemeyer, C.G., Morrison, M.D., Elshahed, M.S., 2012. A critical assessment of the efficacy of biocides used during the hydraulic fracturing process in shale natural gas wells. Int. Biodeterioration Biodegrad. 71, 15–21. http://dx.doi.org/10.1016/j.ibiod.2012.01.013.

Sunde, E., Thorstenson, T., Torsvik, T., 1990. Growth of bacteria on water injection additives. In: Proceedings Volume. 65th Annual SPE Technical Conference, New Orleans, 23–26 September 1990, pp. 727–733.

Sunde, E., Thorstenson, T., Torsvik, T., Vaag, J.E., Espedal, M.S., 1993. Field-related mathematical model to predict and reduce reservoir souring. In: Proceedings Volume. SPE Oilfield Chemistry International Symposium, New Orleans, 2–5 March 1993, pp. 449–456.

Youssef, N., Elshahed, M.S., McInerney, M.J., 2009. Microbial processes in oil fields: Culprits, problems, and opportunities. In: Laskin, A.I., Sariaslani, S., Gadd, G.M. (Eds.), Advances in Applied Microbiology, vol. 66. Academic Press, pp. 141–251 (Chapter 6). <http://www.sciencedirect.com/science/article/B7CSY-4VJJBC8-8/2/44cedf00a850520025373a17cf060e52>.

Zhou, Y., 1990. Bactericide for drilling fluid. Drilling Fluid and Completion Fluid 7 (3), 2A,10–12.

Chapter 18

Proppants

Generally, in a fracturing fluid proppant particles are suspended that are to be placed in the fractures to prevent the fractures from fully closing once the hydraulic pressure is released, thereby forming conductive channels within the formation through which hydrocarbons can flow. Once at least one fracture is created and at least a portion of the proppant is substantially in place, the viscosity of the fracturing fluid may be reduced, to be removed from the formation (Todd et al., 2006).

FLUID LOSS

In certain circumstances, a portion of the fracturing fluid may be lost during the fracturing operation, e.g., through undesirable leakoff into natural fractures present in the formation. This is problematic because such natural fractures often have higher stresses than fractures created by a fracturing operation. These higher stresses may damage the proppant and cause it to form an impermeable plug in the natural fractures, that may prevent hydrocarbons from flowing through the natural fractures.

Conventionally, operators have attempted to solve this problem by including a fluid loss control additive in the fracturing fluid. Conventional fluid loss control additives generally comprise rigid particles having a spheroid shape. The use of these additives can be problematic, because such additives may require particles that have a distinct particle size distribution to achieve efficient fluid loss control.

For example, when such additives are used to block the pore throats in the formation, a sufficient portion of relatively large particles will be required to obstruct the majority of the pore throat. A sufficient portion of relatively small particles will also be required to obstruct the interstices between the large particles. Furthermore, for certain conventional fluid loss control additives, such a desired particle size distribution may be difficult to obtain without incurring the added expense of reprocessing the materials, for example, by cryogenically grinding them to achieve the desired particle size distribution (Todd et al., 2006).

Hydraulic Fracturing Chemicals and Fluids Technology. http://dx.doi.org/10.1016/B978-0-12-411491-3.00018-2
© 2013 Elsevier Inc. All rights reserved.

TRACERS

Fracturing tracers have been originally developed to understand the dynamics of placement, subsequent fluid flowback, and proppant bed cleanup (Asadi et al., 2008).

Nonradioactive chemical tracers have been successfully used to evaluate the behavior. Several classes of chemical compounds were identified that can be placed in diverse sections of the fracturing fluid. In this way, the flowback efficiency of each fluid segment can be assessed.

It is assumed that the relative cleanup of individual fracturing treatment segments in a multiple stage completion process can be monitored. From these data the lateral placement effectiveness of proppants and the vertical communication between zones can be elucidated.

The compatibility of chemical tracers has been investigated, by rheological measurements. As exemplary fluids, a zirconate crosslinked carboxymethyl hydroxypropyl guar-based fluid and a borate crosslinked guar gel were chosen (Sullivan et al., 2004).

A model with an integrated optimization tool with design constraints, design variables, fracture geometry, a production module, and a cost module has been developed. This integrated model is useful for a low permeable oil reservoir (Asadi et al., 2008).

PROPPANT DIAGENESIS

A number of mechanisms have been identified that can degrade the fracture conductivity, including mechanical failure of the proppant grains, liberation of formation fines, proppant embedment, formation spalling, damage from the fracturing fluid, stress cycling, asphaltene deposition, or proppant dissolution (Duenckel et al., 2012). In combination, these factors can reduce the effective conductivity by orders of magnitude as compared with the typically published conductivity data measured under reference conditions.

Another mechanism for the degradation of proppant over time has recently been postulated. This mechanism has been addressed as diagenesis and refers to a dissolution and re-precipitation process that may reduce the porosity, permeability, and strength of the proppant pack during the deposition.

The occurrence and the degree of diagenesis is controlled by closure stress, reservoir temperature, proppant type, and mineralogy of the rock formation. There remains some uncertainty as to predict the occurrence of diagenesis.

Results have been reported from high-temperature static tests, extended-duration flow tests at reservoir conditions, detailed analysis of precipitates, the effects of this environment on the mechanical properties of proppants, chemical and mineralogical analysis of various reservoir shales, and the evaluation of actual proppant samples that have been retrieved from producing wells (Duenckel et al., 2012).

Crystalline precipitates can be formed on the surface of proppants. This is sound for all proppant types, including ceramic, sand, resin-coated materials, and even on inert steel balls or glass rods.

Also zeolites can be formed in the absence of alumina. A careful simulation of reservoir conditions with actual reservoir shale core samples indicates that if diagenetic precipitation does occur, it does not appreciably affect the conductivity performance in the flowing conditions evaluated. Furthermore, other conditions known to occur in oil and gas reservoirs would naturally prevent the formation of zeolites. The results of this work will aid the stimulation engineer in proppant selection and treatment design. While there are many mechanisms contributing to the degradation of proppant performance, these studies indicate that zeolite precipitation is unlikely to be a dominant concern in most wells (Duenckel et al., 2012).

PROPPING AGENTS

For worthwhile oil or gas well stimulation, the best proppant and fluids have to be combined with a good design plan and the right equipment. The selection of a proppant is an important factor in determining how successful the stimulation treatment can be. To select the best proppant for each well, a general understanding of available proppants is imperative.

The propping agents should have high permeability at the respective formation pressures, high resistance to compression, low density, and good resistance to acids. Some propping agents are listed in Table 18.1.

TABLE 18.1 Basic Propping Materials

Material	Description/Property	References
Bauxite	Standard	Andrews (1987) and Fitzgibbon (1986)
Bauxite + ZrO_2	Stress corrosion resistant	Khaund (1987b)
Sand	Low permeability at higher pressures	
Light weights	Specific gravity control	Bienvenu (1996)
Ceramic	Can be fabricated as spheroids	Gibb et al. (1990)
Clay		Fitzgibbon (1988,1989) and Khaund (1987a)

Sand

Sand is the simplest proppant material. Sand is cheap, but at higher stresses it shows a comparatively strong reduction in permeability.

Ceramic Particles

Fired ceramic spheroids have been described for use as a well proppant (Laird and Beck, 1989). Each spheroid has a core made from raw materials comprising mineral particulates, silicium carbide, and a binder. The mixture includes a mineral with chemically bound water or sulfur, which blows the mixture during firing.

Therefore, the core has a number of closed air cells. Each spheroid has an outer shell surrounding the core, comprising a metal oxide selected from aluminum oxide and magnesium oxide. The fired ceramic spheroids have a density of $<2.2\,g\,cm^{-3}$.

Bauxite

Sintered bauxite spheres containing silica are standard proppant materials. The particles have a size range from 0.02 μm to 0.3 μm. They are enhanced to resist stress corrosion by inclusion of 2% zirconia in the mix before firing. A process for manufacturing a material suitable for use as a proppant is characterized by the following steps (Andrews, 1988):

Preparation 18.1. A fine fraction is separated from a naturally occurring bauxite. The fine fraction is an uncalcinated natural bauxite fraction composed largely of monomineralic particles of gibbsite, boehmite, and kaolinite. The kaolinite represents no more than 25% of the total. The separated fine fraction is pelletized in the presence of water. The pellets produced are treated to remove water. ■

Light-weight Proppants

Light-weight propping agents have a specific gravity of less than $2.60\,g\,cm^{-3}$. They are made from kaolin clay and with a light-weight aggregate. Special conditions of calcination are necessary (Lemieux and Rumpf (1990, 1993, 1994)). The alumina content is between 25% and 40% (Sweet, 1993). A high-strength proppant has been described (Bienvenu, 1996) with a specific gravity of $<1.3\,g\,cm^{-3}$.

Ultra light-weight proppants provide excellent transport properties in conventional fracturing fluids with minimal viscosity, which ensures desired effective propped-fracture conductivity (Cawiezel and Gupta, 2010). The use of these ultra light-weight proppants in foamed viscoelastic fluids provides fracturing treatments with optimum proppant placement and excellent cleanup.

Of course, the fluid systems must be optimized. These combined systems require a significant laboratory testing to characterize and optimize the fluid system successfully for the demands of ultralow permeability reservoirs.

Proppants have been described made from sillimanites (Windebank et al., 2010). The sillimanite minerals may be selected from the group consisting of kyanite, sillimanite, and andalusite.

Computational and Experimental Approach

A computational and experimental assessment of light-weight proppants has been undertaken to identify their effectiveness and efficiency to replace sand in hydraulic fracturing treatments in oil or gas well operations (Kulkarni and Ochoa, 2012).

A mixture of ground-nut-shells, aluminum, or ceramic particles can reduce the viscosity of the fracturing fluid while increasing its resistance to compression. Dynamic finite element analysis has been implemented to study the quasi-static compression of a proppant pack, where each particle is modeled individually.

Various mixtures of hard and soft particles have been investigated as a function of shape, size, and inter-particle friction. The particle interactions clearly illustrate changes in pore space as a function of pressure, mixture composition, and friction.

Friction results in a higher porosity by limiting the rearrangement of the particles. The models show that a softer rock with a mixture of hard and soft particles inhibit the flowback, but may decrease the permeability of the pack (Kulkarni and Ochoa, 2012).

Inverted Proppant Convection

In mature fields, it is a continuous challenge to maximize the hydrocarbon recovery while minimizing the associated water production. Water production causes several problems, including scaling, fines migration or sand-face failure, tubular corrosion, and increased hydrostatic loadings (dos Santos et al., 2009).

Thus, although water production is almost an inevitable consequence of oil production, it is usually desirable to defer its onset, or its rise, as long as possible. Proper stimulation is required to prove many reservoirs commercially, including dirty sandstones and lower-permeability layered formations in waterdrive reservoirs or with nearby water zones. The focus on water avoidance has made conformance fracturing an interesting prospect in mature fields because it combines synergistically a relative permeability modifier with a fracturing fluid to enhance production and reduce water cut in one step.

However, if a water zone is below the zone being fractured, fracture invasion may create a conductive path for water production. For example, proppant convection and settling can result in heavier treatment stages displacing rapidly downward from the perforations to the bottom of the fracture. This may occur when treatments call for large pad volumes, high proppant concentrations, or stage density contrasts. An important technique used to avoid this problem is known as inverted proppant convection. It requires proppant buoyancy in the selected fracturing fluid, as is possible when using ultra light-weight proppants with specific gravities from $1.054\,\mathrm{g\,cm^{-3}}to1.75\,\mathrm{g\,cm^{-3}}$.

The technique involves pumping a high-density fluid pad with a specific gravity higher than that of the proppant carrier fluid, which in turn has a specific gravity slightly higher than that of the selected ultra light-weight proppant (dos Santos et al., 2009).

Porous Pack with Fibers

It is possible to build within the formation a porous pack that is a mixture of fibers and the proppant. The fibrous material may be any suitable material, such as natural or synthetic organic fibers, glass fibers, ceramic fibers, carbon fibers.

A porous pack filters out unwanted particles, proppant, and fines, while still allowing production of oil. Using fibers to make a porous pack of fibers and a proppant within the formation reduces the energy consumption of equipment. Pumping the fibers together with the proppant provides significant reductions in the frictional forces that otherwise limit the pumping of fluids containing a proppant (Card et al., 2001).

Coated Proppants

Typically, particulates, such as graded sand, suspended in a portion of the fracturing fluid are then deposited in the fractures when the fracturing fluid is converted to a thin fluid to be returned to the surface. These particulate solids, or proppant particulates, serve to prevent the fractures from fully closing so that conductive channels are formed through which produced hydrocarbons can flow (Dusterhoft et al., 2008).

To prevent the subsequent flowback of the proppant and other particulates with the produced fluids, the proppant may be coated with a curable resin or tackifying agent that may facilitate the consolidation of the proppant particles in the fracture. The partially closed fractures apply pressure to the coated proppant particulates whereby the particulates are forced into contact with each other while the resin or tackifying agent enhances the grain-to-grain contact between individual proppant particles.

The action of the pressure with the tackifying agent ensures the consolidation of the proppant particles into a permeable mass having compressive and tensile strength, while allowing small amounts of deformation at the surface of the proppant packs to reduce the effects of point loading or to reduce proppant crushing (Dusterhoft et al., 2008).

An epoxy resin composition typically includes an oligomeric bisphenol-A epichlorohydrin resin, a 4,4′-diaminodiphenyl sulfone curing agent, a solvent, a silane coupling agent, and a surfactant (Nguyen et al., 2007).

In a series of experiments, three different types of proppant particulates were assessed using a two-component high-temperature epoxy resin system (Dusterhoft et al., 2008). In each experiment, the resin was used in amounts of 3%. The proppant particulate types included bauxite, an intermediate strength proppant, and a light-weight proppant. These proppants are known to withstand pressures from 40 MPa to 80 MPa without substantial crushing. The test temperature was 120 °C for all tests.

The stress was continuously increased over a period of several days from 14 MPa to 80 MPa. Uncoated proppant particulates were also tested. The effects of the closure stresses and flow rates on resin-treated proppant were evaluated with an API linear conductivity cell. The conductivity and permeability of each proppant pack were continuously monitored at 14 MPa (2000 psi) and 120 °C (250 °F) for at least of 25–30 h.

For all the three proppants, there was a significant increase in fracture conductivity and proppant pack permeability for the coated proppants in comparison to the uncoated proppants. The improvement in fracture conductivity and proppant pack permeability was pronounced under lower stress conditions. With coated proppants, there was an evidence that a much more stable interface was created between the proppant and the formation material.

The propping particles can be individually coated with a curable thermoset coating. The coating enhances the chemical resistance of the proppants.

This modification is necessary if a proppant is not stable against the additives in the fracturing fluid, such as an acid gel breaker. Resole-type phenolic resins are recommended as coating materials in the presence of oxidative gel breakers (Dewprashad, 1995). Polymer coatings for propping agents are listed in Table 18.2.

Another advantage of a coating is the reduction of friction of the proppant particles. In this case, coating materials with low friction coefficients are included into the mixture of the coating (de Grood and Baycroft, 2010). Such materials are summarized in Table 18.3. Of course, not all of the materials listed in Table 18.3 are reasonable from the view of economics. Multiple coating of particulate material results in a final coated product that has a smooth and uniform surface.

Antisettling Additives

Proppant transport inside a hydraulic fracture has two components, when the fracture is being generated. The horizontal component is dictated by the fluid velocity and by the associated streamlines which help to carry the proppant to the tip of the fracture. The vertical component is dictated by the

TABLE 18.2 Polymer Coatings for Propping Agents

Material	References
Phenolic/furan resin or furan resin[a]	Armbruster (1987)
Novolak epoxide resin[a]	Gibb et al. (1989)
Pyrolytic carbon coating[a]	Hudson and Martin (1989)
Bisphenolic resin[b]	
Phenolic resin	Johnson et al. (1993)
Furfuryl alcohol resin[b]	Ellis and Surles (1997)
Bisphenol-A resin (curable)[b]	Johnson and Tse (1996)
Epoxide resin with N-β-(aminoethyl)-δ-aminopropyltrimethoxysilane crosslinker	Nguyen et al. (2001)
Poly(amide) and others	Nguyen and Weaver (2001)

[a]*Chemically resistant.*
[b]*Flowback prevention.*

TABLE 18.3 Friction Reducing Materials (de Grood and Baycroft, 2010)

Material	Material
Antimony trioxide	Bismuth
Boric acid	Calcium barium fluoride
Copper	Graphite
Indium	Lead oxide
Lead sulfide	Molybdenum disulfide
Niobium diselenide	Fluoropolymers
Poly(tetrafluoroethylene)	Silver
Tin	Tungsten disulfide

terminal particle settling velocity of the particle and is a function of proppant diameter and density as well as fluid viscosity and density (Watters et al., 2010).

By infusing an additive that has a lower density than the fluid medium, the density gradients inside the fracture can be varied. The upward movement of the low density additive will interfere with the downward movement of high-density proppant and vice versa.

This mutual interference between the proppants and the additive confined in the narrow fracture will significantly hinder the settling of the high-density proppant. By this principle, the times of settling of the proppant can be controlled.

TABLE 18.4 Time of Settling in the Presence of a Low Density Additive (Watters et al., 2010)

Distance (ft)	Time to Settle (s)	
Additive Added →	(0%)	(5%)
1	8	11
2	22	42
3	47	61
4	56	77

The low-density material should have a particle size distribution similar to that of standard proppant. In addition to be a buoyant, the material may also act as a proppant. Examples of such materials are poly(lactic acid) particles or glass beads. The influence of such a low-density additive on the time of settling is summarized in Table 18.4.

The tests were conduced in a $15\,\mathrm{lb\,Mgal^{-1}}$ ($1.8\,\mathrm{g\,m^{-3}}$) guar-based linear gel. The apparent viscosity of the gel was $8\,\mathrm{cP}$ at $511\,\mathrm{s^{-1}}$. The linear gels for testing contained a proppant concentration of $240\,\mathrm{kg\,m^{-3}}$ with a diameter of 30–50 mesh.

Proppant Flowback

The flowback of a proppant following fracture stimulation treatment is a major concern because of the damage to equipment and loss in well production. The mechanisms of flowback and the methods to control flowback have been discussed in the literature (Nguyen et al., 1996b). To reduce the proppant flowback, a curable resin-coated proppant can be applied (Nimerick et al., 1990). The agent must be placed across the producing interval to prevent or at least to reduce the proppant flowback.

Thermoplastic Films

Thermoplastic film materials have been developed to reduce the proppant flowback that can occur after fracturing treatments (Nguyen et al., 1996a,b). A heat-shrinkable film cut into thin slivers provides a flowback reduction over broad temperature ranges and closure stress ranges and was found to cause little impairment to the fracture conductivity, but with some dependency on concentration, temperature, and closure stress.

Adhesive-Coated Material

The addition of an adhesive-coated material to proppants decreases the flowback of the particulates (Caveny et al., 1996). Such adhesive-coated materials can be inorganic or organic fibers. The adhesive-coated material interacts mechanically with the proppant particles to prevent the flowback of particulates to the wellbore.

The consolidation of a proppant also may occur via a poly(urethane) coating, which will slowly polymerize after the fracturing treatment due to a polyaddition process (Wiser-Halladay, 1990).

Magnetized Material

A magnetized material in the form of beads, fibers, strips, or particles can be placed together with a proppant. The magnetized material moves to voids or channels located within the proppant bed and forms clusters, which are held together by magnetic attraction in the voids or channels, which in turn facilitate the formation of permeable proppant bridges therein.

The magnetized material-proppant bridges retard and ultimately prevent the flowback of proppant and formation solids, but still allow the production of oil and gas through the fracture at sufficiently high rates (Clark et al., 2000). In a similar way, fibrous bundles placed with the proppant may act in a fracture as a flowback preventer (Nguyen and Schreiner, 1999).

Tradenames in References

Tradename	Supplier
Description	
SandWedge® NT	Halliburton Energy Services
Tackifying compound, based on 2-methoxymethylethoxy propanol (Nguyen et al., 2007)	

REFERENCES

Andrews, W.H., 1987. Bauxite proppant. US Patent 4713203, assigned to Comalco Aluminium Ltd., Victoria, AU, 15 December 1987. <http://www.freepatentsonline.com/4713203.html>.

Andrews, W.H., 1988. Sintered bauxite pellets and their application as proppants in hydraulic fracturing. AU Patent 579242, 17 November 1988.

Armbruster, D.R., 1987. Precured coated particulate material. US Patent 4694905, 22 September 1987.

Asadi, M., Woodroof, R.A., Himes, R.E., 2008. Comparative study of flowback analysis using polymer concentrations and fracturing-fluid tracer methods: a field study. SPE Prod. Oper. 23 (2), 147–157. http://dx.doi.org/10.2118/101614-PA.

Bienvenu Jr., R.L., 1996. Lightweight proppants and their use in hydraulic fracturing. US Patent 5531274, 2 July 1996.

Card, R.J., Howard, P.R., Feraud, J.P., Constien, V.G., 2001. Control of particulate flowback in subterranean wells. US Patent 6172011, assigned to Schlumberger Technol. Corp., 9 January 2001.

Caveny, W.J., Weaver, J.D., Nguyen, P.D., 1996. Control of particulate flowback in subterranean wells. US Patent 5 582 249, 10 December 1996.
Cawiezel, K.E., Gupta, D.V.S., 2010. Successful optimization of viscoelastic foamed fracturing fluids with ultralightweight proppants for ultralow-permeability reservoirs. SPE Prod. Oper. 25 (1), 80–88. http://dx.doi.org/10.2118/119626-PA.
Clark, M.D., Walker, P.L., Schreiner, K.L., Nguyen, P.D., 2000. Methods of preventing well fracture proppant flowback. US Patent 6 116 342, assigned to Halliburton Energy Service, 12 September 2000.
de Grood, R.J.C., Baycroft, P.D., 2010. Use of coated proppant to minimize abrasive erosion in high rate fracturing operations. US Patent 7 730 948, assigned to Baker Hughes Incorp., Houston, TX, 8 June 2010. <http://www.freepatentsonline.com/7730948.html>.
Dewprashad, B., 1995. Method of producing coated proppants compatible with oxidizing gel breakers. US Patent 5 420 174, assigned to Halliburton Co., 30 May 1995.
dos Santos, J.A.C.M., Cunha, R.A., de Melo, R.C.B., Aboud, R.S., Pedrosa, H.A., Marchi, F.A., 2009. Inverted-convection proppant transport for effective conformance fracturing. SPE Prod. Oper. 24 (1), 187–193. http://dx.doi.org/10.2118/109585-PA.
Duenckel, R., Conway, M.W., Eldred, B., Vincent, M.C., 2012. Proppant diagenesis-integrated analyses provide new insights into origin, occurrence, and implications for proppant performance. SPE Prod. Oper. 27 (2), 131–144.
Dusterhoft, R.G., Fitzpatrick, H.J., Adams, D., Glover, W.F., Nguyen, P.D., 2008. Methods of stabilizing surfaces of subterranean formations. US Patent 7 343 973, assigned to Halliburton Energy Services, Inc., Duncan, OK, 18 March 2008. <http://www.freepatentsonline.com/7343973.html>.
Ellis, P.D., Surles, B.W., 1997. Chemically inert resin coated proppant system for control of proppant flowback in hydraulically fractured wells. US Patent 5 604 184, assigned to Texaco Inc., 18 February 1997.
Fitzgibbon, J.J., 1986. Use of uncalcined/partially calcined ingredients in the manufacture of sintered pellets useful for gas and oil well proppants. US Patent 4 623 630, 18 November 1986.
Fitzgibbon, J.J., 1988. Sintered, spherical, composite pellets prepared from clay as a major ingredient useful for oil and gas well proppants. CA Patent 1 232 751, 16 February 1988.
Fitzgibbon, J.J., 1989. Sintered spherical pellets containing clay as a major component useful for gas and oil well proppants. US Patent 4 879 181, 7 November 1989.
Gibb, J.L., Laird, J.A., Berntson, L.G., 1989. Novolac coated ceramic particulate. EP Patent 308 257, 22 March 1989.
Gibb, J.L., Laird, J.A., Lee, G.W., Whitcomb, W.C., 1990. Particulate ceramic useful as a proppant. US Patent 4 944 905, 31 July 1990.
Hudson, T.E., Martin, J.W., 1989. Pyrolytic carbon coating of media improves gravel packing and fracturing capabilities. US Patent 4 796 701, 10 January 1989.
Johnson, C.K., Tse, K.T., 1996. Bisphenol-containing resin coating articles and methods of using same. EP Patent 735 234, assigned to Borden Inc., 2 October 1996.
Johnson, C.R., Tse, K.T., Korpics, C.J., 1993. Phenolic resin coated proppants with reduced hydraulic fluid interaction. US Patent 5 218 038, assigned to Borden, Inc., Columbus, OH, 8 June 1993. <http://www.freepatentsonline.com/5218038.html>.
Khaund, A., 1987a. Sintered low density gas and oil well proppants from a low cost unblended clay material of selected composition. US Patent 4 668 645, 26 May 1987.
Khaund, A.K., 1987b. Stress-corrosion resistant proppant for oil and gas wells. US Patent 4 639 427, 27 January 1987.
Kulkarni, M.C., Ochoa, O.O., 2012. Mechanics of light weight proppants: a discrete approach. Compos. Sci. Technol. 72 (8), 879–885. http://dx.doi.org/10.1016/j.compscitech.2012.02.017.
Laird, J.A., Beck, W.R., 1989. Ceramic spheroids having low density and high crush resistance. EP Patent 207 668, 5 April 1989.
Lemieux, P.R., Rumpf, D.S., 1990. Low density proppant and methods for making and using same. EP Patent 353 740, 7 February 1990.
Lemieux, P.R., Rumpf, D.S., 1993. Lightweight oil and gas well proppant and methods for making and using same. AU Patent 637 576, 3 June 1993.
Lemieux, P.R., Rumpf, D.S., 1994. Lightweight proppants for oil and gas wells and methods for making and using same. CA Patent 1 330 255, 21 June 1994.

Nguyen, P.D., Weaver, J.D., Parker, M.A., King, D.G., 1996a. Thermoplastic film prevents proppant flowback. Oil Gas J. 94 (6), 60–62.

Nguyen, P.D., Weaver, J.D., Parker, M.A., King, D.G., Gillstrom, R.L., Van Batenburg, D.W., 1996b. Proppant flowback control additives. In: Proceedings Volume. Annual SPE Technical Conference, Denver, 6–9 October 1996, pp. 119–131.

Nguyen, P.D., Schreiner, K.L., 1999. Preventing well fracture proppant flow-back. US Patent 5 908 073, assigned to Halliburton Energy Serv., 1 June 1999.

Nguyen, P.D., Weaver, J.D., 2001. Method of controlling particulate flowback in subterranean wells and introducing treatment chemicals. US Patent 6 209 643, assigned to Halliburton Energy Serv., 3 April 2001.

Nguyen, P.D., Weaver, J.D., Brumley, J.L., 2001. Stimulating fluid production from unconsolidated formations. US Patent 6 257 335, assigned to Halliburton Energy Serv., 10 July 2001.

Nguyen, P.D., Barton, J.A., Isenberg, O.M., 2007. Methods and compositions for consolidating proppant in fractures. US Patent 7 264 052, assigned to Halliburton Energy Services, Inc., Duncan, OK, 4 September 2007. <http://www.freepatentsonline.com/7264052.html>.

Nimerick, K.H., McConnell, S.B., Samuelson, M.L., 1990. Compatibility of resin-coated proppants with crosslinked fracturing fluids. In: Proceedings Volume. 65th Annual SPE Technical Conference, New Orleans, 23–26 September 1990, pp. 245–250.

Sullivan, R., Woodroof, R., Steinberger-Glaser, A., Fielder, R., Asadi, M., 2004. Optimizing fracturing fluid cleanup in the Bossier sand using chemical frac tracers and aggressive gel breaker deployment. In: Proceedings of SPE Annual Technical Conference and Exhibition. Society of Petroleum Engineers. http://dx.doi.org/10.2118/90030-MS.

Sweet, L., 1993. Method of fracturing a subterranean formation with a lightweight propping agent. US Patent 5 188 175, 23 February 1993.

Todd, B.L., Slabaugh, B.F., Munoz Jr., T., Parker, M.A., 2006. Fluid loss control additives for use in fracturing subterranean formations. US Patent 7 096 947, assigned to Halliburton Energy Services, Inc., Duncan, OK, 29 August 2006. <http://www.freepatentsonline.com/7096947.html>.

Watters, J.T., Ammachathram, M., Watters, L.T., 2010. Method to enhance proppant conductivity from hydraulically fractured wells. US Patent 7 708 069, assigned to Superior Energy Services, L.L.C., New Orleans, LA, 4 May 2010. <http://www.freepatentsonline.com/7708069.html>.

Windebank, M., Hart, J., Alary, J.A., 2010. Proppants and anti-flowback additives made from sillimanite minerals, methods of manufacture, and methods of use. US Patent 7 790 656, assigned to Imerys, Paris, FR, 7 September 2010. <http://www.freepatentsonline.com/7790656.html>.

Wiser-Halladay, R., 1990. Polyurethane quasi prepolymer for proppant consolidation. US Patent 4 920 192, 24 April 1990.

Chapter 19

Special Compositions

Subsequently some special compositions for fracturing fluids with exceptional performance are reviewed.

HEAT-GENERATING SYSTEM

During the fracturing treatment for low temperature, shallow and high freezing point oil reservoirs, the major problems are to overcome uncompleted breakdown, uncompleted cleanup of fracturing fluids, and cold damages to the formations by injecting cold fluid.

To avoid those problems, an encapsulated heat-generating hydraulic fracturing fluid system has been developed. Two kinds of heat-generating systems are commonly used in the production of oil and gas. In a hydrogen peroxide-based system, the reaction occurs as (Bayless, 2000):

$$H_2O_2 \rightarrow \frac{1}{2}O_2 + H_2O. \tag{19.1}$$

The standard heat of reaction of this system, ΔH is $-196\,\text{kJ mol}^{-1}$. In a nitrite ammonium salt system the heat-generating reaction is as (Wu et al., 2005):

$$NO_2^- + NH_4^+ \rightarrow N_2 + H_2O. \tag{19.2}$$

The standard heat of reaction of this system, ΔH is $-333\,\text{kJ mol}^{-1}$, i.e., much more than of the first system.

The second system was investigated closely, and oxalic acid was chosen as a catalyst of reaction and encapsulated using ethyl cellulose and paraffin as coating materials by the phase separation method. The components and cyclohexane were added, heated to 81 °C under continuing stirring until the ethyl cellulose dissolved. After a stirring time of 30 min and cooling, the intermediate products were recovered by filtering, washed with cyclohexane, and dried.

The reaction kinetics of this system can be described as

$$\frac{dc}{dt} = -1.267 \times 10^7 C_H^{1.17} C_0^{1.88} \exp\left(-\frac{5360}{T}\right). \tag{19.3}$$

Hydraulic Fracturing Chemicals and Fluids Technology. http://dx.doi.org/10.1016/B978-0-12-411491-3.00019-4
© 2013 Elsevier Inc. All rights reserved.

TABLE 19.1 Fracturing Fluid Composition (Wu et al., 2005)

Component	Amount (%)
Hydroxypropyl guar	0.6
Borax	0.7
Ammonium per sulfate	0.08
Heat-generating agent	**Amount (mol l^{-1})**
Experiment 1	1.5
Experiment 2	1.75
Experiment 3	2.0

Here, $\frac{dc}{dt}$ is the rate of consumption of the reactants in $\mathrm{mol\,l^{-1}\,min^{-1}}$, C_H is the concentration of the proton catalyst, and C_0 is the initial concentration, all in $\mathrm{mol\,l^{-1}}$. The expression closes with an Arrhenius factor.

A fracturing fluid composition has been formulated as shown in Table 19.1. The results showed that a hydraulic fracturing fluid containing encapsulated heat-generating agents has good stability and compatibility.

When the fracturing fluid contains 2.0 mol l^{-1} heat-generating agent 0.93% encapsulated oxalic acid and 0.08% ammonium persulfate, the peak temperature can reach 78 °C and the viscosity of residual liquid is 3.12 MPa s after 4 h (Wu et al., 2005).

CROSSLINKABLE SYNTHETIC POLYMERS

A hydraulic fracturing fluid has been developed that is capable of reaching fluid service temperatures up to 232 °C (Holtsclaw and Funkhouser, 2010). This fluid technology uses a synthetic polymer that is crosslinkable with metal ions to generate high viscosity. The polymer-based fracturing gel overcomes the thermal limitations of traditional guar and derivatized guar-based fracturing fluids. An acrylamide terpolymer based on acrylic acid, acrylamide, and 2-acrylamido-2-methyl-1-propane sulfonic acid was used.

The efforts in research resulted in a fracturing fluid with a good fluid stability at high temperatures to create better proppant transport and placement in these most-demanding environments. The crosslinking reaction can be tuned for the onset of crosslinking from 38 °C to 138 °C, thus allowing the optimization for special well conditions.

Rheological data demonstrate the fluid stability, crosslinking performance, and controlled fluid breaks. In addition, dynamic fluid loss and regained-conductivity data have been presented to illustrate the fluid cleanup in the proppant packs (Holtsclaw and Funkhouser, 2010).

SINGLE PHASE MICROEMULSION

A fracturing fluid based on a combination of a single phase microemulsion and a gelable polymer system has been developed (Liu et al., 2010). This formulation is suitable to reduce the polymer loading. Also other properties of the gelled polymer are improved with a good synergistic effect.

The formulation of a single phase microemulsion system was obtained from data of the corresponding phase diagram. The formulations were prepared by adding a single phase microemulsion to a gelable polymer system at various concentrations with the characteristics of high viscosity, low fluid loss, and low friction.

Furthermore, the gel systems have low residues remaining in formation, low surface tension, low pressure to initiate cleanup, and high core permeability maintaining. The formulation is expected to have the advantages of reducing formation damage, lowering the initiate cleanup pressure, and maintaining the original core regain for fracturing treatment.

CROSSLINKING COMPOSITION

A composition suitable for crosslinking has been described that is based on a triethanolamine complex of zirconium, tetra triethanolamine zirconate, and tetra(hydroxyalkyl)ethylenediamine in water (Putzig, 2010a).

Tetra triethanolamine zirconate can be synthesized by the reaction of tetra *n*-propyl zirconate with triethanolamine in an alcohol solution. Afterwards, lactic acid is added to get a zirconium-alkanolamine-hydroxycarboxylic acid complex. The method of preparation has been described in detail (Putzig, 2010b).

REFERENCES

Bayless, J.H., 2000. Hydrogen peroxide applications for the oil industry. World Oil 221 (5), 50–54.

Holtsclaw, J., Funkhouser, G.P., 2010. A crosslinkable synthetic-polymer system for high-temperature hydraulic-fracturing applications. SPE Drill. Completion 25 (4), 555–563. <http://www.onepetro.org/mslib/servlet/onepetropreview?id=SPE-125250-PA>.

Liu, D.-X., Fan, M.-F., Yao, L.-T., Zhao, X.-T., Wang, Y.-L., 2010. A new fracturing fluid with combination of single phase microemulsion and gelable polymer system. J. Petrol. Sci. Eng. 73 (3–4), 267–271. http://dx.doi.org/10.1016/j.petrol.2010.07.008.

Putzig, D.E., 2010a. Hydraulic fracturing methods using cross-linking composition comprising zirconium triethanolamine complex. US Patent 7 730 952, assigned to E.I. duPont de Nemours and Co., Wilmington, DE, 8 June 2010. <http://www.freepatentsonline.com/7730952.html>.

Putzig, D.E., 2010b. Process to prepare zirconium-based cross-linker compositions and their use in oil field applications. US Patent 7 754 660, assigned to E.I. duPont de Nemours and Co., Wilmington, DE, 13 July 2010. <http://www.freepatentsonline.com/7754660.html>.

Wu, J., Zhang, N., Wu, X., Liu, X., 2005. Experimental research on a new encapsulated heat-generating hydraulic fracturing fluid system. Chin. J. Geochem. 25 (2), 162–166. http://dx.doi.org/10.1007/BF02872176.

Chapter 20

Environmental Aspects

RISK ANALYSIS

In recent years, shale gas formations have become economically viable through the use of horizontal drilling and hydraulic fracturing. These techniques bear potential environmental risks due to high amounts of water in use and are therefore a substantial risk for water pollution.

By probability methods, the likelihood of a water contamination from natural gas extraction in the Marcellus Shale was assessed (Rozell and Reaven, 2012). The Marcellus Shale formation is located in northeastern United States in the south of Erie lake and Ontario lake.

Probability bounds analysis is well suited when data are sparse and the parameters highly uncertain. Five pathways of water contamination could be identified (Rozell and Reaven, 2012):

1. Transportation spills,
2. Well casing leaks,
3. Leaks through fractured rock,
4. Drilling site discharge, and
5. Wastewater disposal.

Probability boxes have been generated for each pathway. The potential contamination risk and epistemic uncertainty associated with hydraulic fracturing wastewater disposal was several orders of magnitude larger than the other pathways.

Even in a best-case scenario, it was very likely that an individual well would release at least 200 m^3 of contaminated fluids. Because the total number of wells in the Marcellus Shale region could range into the tens of thousands, this substantial potential risk suggested that additional steps should be taken to reduce the potential for contaminated fluid leaks. To reduce the considerable epistemic uncertainty, more data should be collected on the ability of industrial and municipal wastewater treatment facilities to remove contaminants from a used hydraulic fracturing fluid (Rozell and Reaven, 2012).

Hydraulic Fracturing Chemicals and Fluids Technology. http://dx.doi.org/10.1016/B978-0-12-411491-3.00020-0
© 2013 Elsevier Inc. All rights reserved.

Mass balance principles were applied to a four-compartment partition model for 12 different hazardous components of hydraulic fracturing fluid additives used in 47 completed natural gas wells in the Marcellus Shale. Spill scenarios were modeled as if 1000 gal of diluted additive were discharged into a surface water body or onto soil. The resulting concentrations were ranked according to magnitude, providing a relative comparison of quantities to be expected in each compartment. The highest mass concentrations in the water, soil, and biota compartments were due to sodium hydroxide, 4,4-dimethyl oxazolidine, and hydrochloric acid.

4,4-Dimethyl oxazolidine ranked highest in the air compartment (Aminto and Olson, 2012).

CONTAMINATED WATER RECLAIM

Systems and methods have been developed for reclaiming water containing contaminants typically associated with produced water in order to produce a treated water having a quality adequate for reuse as a fracturing water (Shafer et al., 2009).

Produced water includes natural contaminants that come from the subsurface environment, such as hydrocarbons from the oil-bearing or gas-bearing strata and inorganic salts. The produced water may also include man-made contaminants, that may contain spent fracturing fluids including polymers and inorganic crosslinking agents, polymer breaking agents, friction-reduction chemicals, and artificial lubricants. A special problem occurs when the produced water is additionally contaminated with methanol. Namely, methanol will pass through nearly any available membrane filtration systems.

A system for contaminated water reclaiming includes anaerobically digesting the contaminated water, followed by aerating the water to enhance biological digestion. After aeration, the water is separated using a flotation operation that effectively removes the spent friction-reducing agents and allows the treated water to be reclaimed and reused as fracturing water (Shafer et al., 2009). In a separate branch of the unit, after a sand pack filter a series of bioreactors follows and finally a boron treatment unit. This so cleaned waste can be discharged to the environment.

The typical and target values of contaminant concentrations obtained from such a system are shown in Table 20.1.

Wastewater from Hydro-fracturing

Hydro-fracturing can generate large amounts of wastewater. The well-drilling process involves injecting water, along with sand and a mixture of chemicals. The process involves pumping water into fractures at pressures exceeding 200 bar and flow rates exceeding $5\,l\,s^{-1}$ in order to create a long fracture sand

TABLE 20.1 Contaminant Concentrations (Shafer et al., 2009)

Contaminant	Typical Range ($mg\,ml^{-1}$)	Target (maximum) ($mg\,ml^{-1}$)
Total dissolved solids 180 °C	9000–16,000	10,000
Total dissolved solids 105 °C	0–100	75
Total organic carbon	400–800	700
Chemical oxygen demand	1000–3000	2000
Biological oxygen demand	500–1500	1000
Iron	1–10	5
Chloride	5000–10,000	6000
Potassium	100–500	300
Calcium	50–250	150
Magnesium	10–100	25
Sodium	2000–5000	3000
Sulfate	40–200	50
Carbonate	0–100	25
Bicarbonate	100–1200	800
Boron	0–20	15

pack intersecting with natural fractures in the shale. In total, the hydro-fracturing process can use 10,000 m^3 of water (DiTommaso and DiTommaso, 2012).

A method for cleaning the post-drilling flowback water from hydro-fracturing has been presented. This method includes (DiTommaso and DiTommaso, 2012):

- Removing the oil from the flowback water.
- Filtering the flowback water using an ultra filter with a pore size of about 0.1 μm or less to remove solid particulates and large organic molecules, such as benzene, ethylbenzene, toluene, and xylene, from the water.
- Concentrating the flowback water to produce a brine that contains from about 15% to 40% of salt relative to the total weight of the flowback brine.
- Performing one or more chemical precipitation processes using an effective amount of reagents to precipitate out the desired high-quality commercial products, such as, barium sulfate, strontium carbonate, or calcium carbonate.
- Crystallizing the chemically treated and concentrated flowback brine to produce pure salt products, such as sodium and calcium chloride.

Phosphorus Recovery in Flowback Fluids

Phosphonate esters are used as gellants in fracturing. In the course of those operations, phosphorus may be included in the crude oil outflow. However, refinery plugging caused by volatile phosphorus components originating from phosphate ester oil gellants has been described which is a potential risk.

A maximum of 0.5 ppm volatile phosphorus in crudes has been proposed to avoid costly unplanned refinery shutdowns. This specification is based on what is considered achievable through a combination of new chemistry and typical field dilution. However, this specification is based on average concentrations of phosphorus added to the oil to gel it and assumes the oil is initially free from phosphorus.

In some flowback studies, total and resulting volatile phosphorus concentrations in a great excess of that added have been observed. Potential explanations for phosphorus concentrations significantly higher than those originally added have been given (Fyten et al., 2007).

An inductively coupled plasma optical emission spectroscopy based method has been developed for the analysis of total volatile phosphorus in distillate fractions of crude oil (Nizio and Harynuk, 2012). However, this method is plagued by poor precision and a high limit of detection of $0.5\ mg\ ml^{-1}$.

Another approach uses comprehensive two-dimensional gas chromatography with a nitrogen phosphorus detector. This method provides qualitative and quantitative profiles of alkyl phosphates in industrial petroleum samples with increased precision. A recovery study in a fracturing fluid sample and a profiling study of alkyl phosphates in four recovered fracturing fluids or flowback crude oil mixtures has been presented (Nizio and Harynuk, 2012).

GREEN FORMULATIONS

Biodegradable Chelants

Today, oil companies are requesting *green* fracturing fluid chemistry. Commonly used demulsifiers and scale inhibitors tend to have environmental concerns and do not have a multifunctional character.

Biodegradable and nontoxic chelant compositions can perform multiple beneficial functions in an aqueous fracturing fluid through the chelation of ions. Some of the multiple functions include various combinations of the following (Crews, 2006):

- Demulsifier,
- Demulsifier enhancer,
- Scale inhibitor,
- Crosslink delay agent,
- Crosslinked gel stabilizer,
- Enzyme breaker, and
- Stabilizer.

TABLE 20.2 Environmental Friendly Crosslinking Agents (Vaughn et al., 2011)

Chemical Description	Tradename
C_6 alcohol zirconium alkoxide	Vertec™ XL985
Tetrakis(2-ethylhexyl) zirconate	
C_4 glycol zirconium alkoxide	Vertec™ XL980
C_6 alcohol titanium alkoxide	Vertec™ XL121
Tetrakis(2-ethylhexyl) titanate	
C_4 glycol titanium alkoxide	Vertec™ XL990

Nontoxic Flowback Formulation

A nontoxic, environmentally friendly, green flowback aid that reduces water blockage when injected into a fractured reservoir has been described (Berger and Berger, 2011).

Preparation 20.1. The flowback aid has been prepared by dissolving 15 parts of ethyl lactate in 15 parts of methyl tallate. Then, a second solution was prepared by mixing 12.5 parts of lauryl alcohol containing 6 mol of ethylene oxide with a 12.5 parts of a 70% aqueous solution of sodium dodecyl sulfosuccinate. Eventually, the two solutions were combined under stirring and 55 parts of water was slowly added to form a clear low-viscosity stable solution. ■

Crosslinking Agents

Group IVB metal alkoxide crosslinkers are not both less flammable and environmentally more acceptable with respect to transportation and environmental concerns (Vaughn et al., 2011). These products are commercially available and listed in Table 20.2.

Several methods of how to prepare crosslinking formulations from these compounds have been described in detail, as well as the rheological performance of the formulations and the degree of flame retardancy (Vaughn et al., 2011).

SELF-DEGRADING FOAMING COMPOSITION

A pH-dependent foamed fracturing fluid has been described. The fracturing fluid is made by combining a gelling agent, a surfactant, and a proppant. The surfactant is capable of facilitating foaming of the fracturing fluid at the initial pH and defoaming of the fracturing fluid when its pH is changed. The pH of the fracturing fluid changes by an in situ contact with an acidic material, causing the level of foam in the fracturing fluid to be reduced. As a result of the reduction

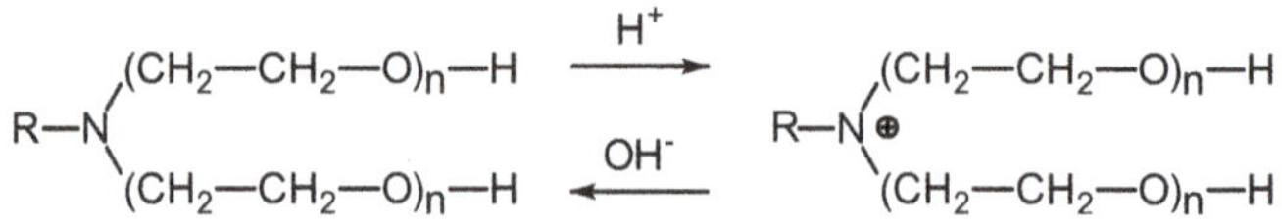

$$R{-}N\begin{matrix}(CH_2{-}CH_2{-}O)_n{-}H\\(CH_2{-}CH_2{-}O)_n{-}H\end{matrix}\ \underset{OH^-}{\overset{H^+}{\rightleftharpoons}}\ R{-}N^{\oplus}\begin{matrix}(CH_2{-}CH_2{-}O)_n{-}H\\(CH_2{-}CH_2{-}O)_n{-}H\end{matrix}$$

FIGURE 20.1 Protonation reaction (Chatterji et al., 2005).

of the foam, the fracturing fluid deposits the proppant into the fractures formed in the subterranean formation (Chatterji et al., 2005). The fracturing fluid may be returned to the surface for recovery. There it is re-foamed by changing its pH back and again injected downhole.

Both an amphoteric surfactant and an anionic surfactant may be used. A suitable surfactant is a tertiary alkyl amine ethoxylate. This compound may be converted from a foaming surfactant to a defoaming surfactant by the addition of hydrogen ions. By the addition of hydroxide the foaming property turns back. The protonation reaction is shown in Figure 20.1.

A self-degrading foaming composition has been described that is a mixture of anionic surfactant and nonionic surfactant. The fracturing fluid forms a substantially less stable foam when the foamed fracturing fluid is recovered during reclaim (Dahanayake et al., 2008).

A preferred anionic surfactant in the self-degrading foaming composition is disodium lauramide monoethanolamine sulfosuccinamate. As nonionic surfactants, ethoxylated fatty acid esters of poly (ethylene glycol) may find use. It has been found that the foam quality degrades more quickly at a higher pH.

Tradenames in References

Tradename	Supplier
Description	
Lewatit® (Series)	Lanxess Deutschland GmbH
Divinylbenzene/styrene copolymer ion exchange resins (Shafer et al., 2009)	
Vertec™ (Seris)	
Environmental friendly crosslinking agents (Vaughn et al., 2011)	

REFERENCES

Aminto, A., Olson, M.S., 2012. Four-compartment partition model of hazardous components in hydraulic fracturing fluid additives. J. Nat. Gas Sci. Eng. 7, 16–21. http://dx.doi.org/10.1016/j.jngse.2012.03.006.

Berger, P.D., Berger, C.H., 2011. Environmental friendly fracturing and stimulation composition and method of using the same. US Patent 7 998 911, assigned to Oil Chem. Tech., Sugar Land, TX, 16 August 2011. <http://www.freepatentsonline.com/7998911.html>.

Chatterji, J., King, K.L., King, B.J., Slabaugh, B.F., 2005. Methods of fracturing a subterranean formation using a pH dependent foamed fracturing fluid. US Patent 6966379, assigned to Halliburton Energy Services, Inc., Duncan, OK, 22 November 2005. <http://www.freepatentsonline.com/6966379.html>.

Crews, J.B., 2006. Biodegradable chelant compositions for fracturing fluid. US Patent 7078370, assigned to Baker Hughes Incorp., Houston, TX, 18 July 2006. <http://www.freepatentsonline.com/7078370.html>.

Dahanayake, M.S., Kesavan, S., Colaco, A., 2008. Method of recycling fracturing fluids using a self-degrading foaming composition. US Patent 7 404 442, assigned to Rhodia Inc. (Cranbury, NJ), 29 July 2008. <http://www.freepatentsonline.com/7404442.html>

DiTommaso, F.A., DiTommaso, P.N., 2012. Method of making pure salt from frac-water/wastewater. US Patent 8273320, assigned to FracPure Holdings LLC, Dover, DE, 25 September 2012. <http://www.freepatentsonline.com/8273320.html>.

Fyten, G., Houle, P., Taylor, R.S., Stemler, P.S., Lemieux, A., 2007. Total phosphorus recovery in flowback fluids after gelled hydrocarbon fracturing fluid treatments. J. Can. Petrol. Technol. 46 (12), 17–21. http://dx.doi.org/10.2118/07-12-TN2.

Nizio, K.D., Harynuk, J.J., 2012. Analysis of alkyl phosphates in petroleum samples by comprehensive two-dimensional gas chromatography with nitrogen phosphorus detection and post-column deans switching. J. Chromatogra. A 1252, 171–176. http://dx.doi.org/10.1016/j.chroma.2012.06.070.

Rozell, D.J., Reaven, S.J., 2012. Water pollution risk associated with natural gas extraction from the Marcellus Shale. Risk Anal. 32 (8), 1382–1393. http://dx.doi.org/10.1111/j.1539-6924.2011.01757.x.

Shafer, L.L., James, J.W., Rath, R.D., Eubank, J., 2009. Method for generating fracturing water. US Patent 7527736, assigned to Anticline Disposal, LLC, Rapid City, SD, 5 May 2009. <http://www.freepatentsonline.com/7527736.html>.

Vaughn, D.E., Duncan, R.H., Harry, D.N., Williams, D.A., 2011. Non-flammable, non-aqueous group ivb metal alkoxide crosslinkers and fracturing fluid compositions incorporating same. US Patent 7879771, assigned to Benchmark Performance Group, Inc., Houston, TX, 1 February 2011. <http://www.freepatentsonline.com/7879771.html>.

Index

TRADENAMES

ACRONYMS

CHEMICALS

Boldface numbers refer to Figures

GENERAL INDEX